Cybersecurity 2050

This book explores the critical intersection of human behavior, cybersecurity, and the transformative potential of quantum technologies. It delves into the vulnerabilities and resilience of human intelligence in the face of cyber threats, examining how cognitive biases, social dynamics, and mental health can be exploited in the digital age.

Cybersecurity 2050: Protecting Humanity in a Hyper-Connected World explores the cutting-edge applications of quantum computing in cybersecurity, discussing the efficiency of quantum security algorithms on Earth and over space communications such as those needed to inhabit Mars. The challenges and opportunities of human life on extraterrestrial worlds, such as Mars, will further shape the evolution of human intelligence. The isolated and confined environment of a Martian habitat, coupled with the reliance on advanced technologies for survival, will demand new forms of adaptability, resilience, and social cooperation. The author addresses the imminent revolution in cybersecurity regulations and transforms the attention of the bright minds of businesses and policymakers for the challenges and opportunities of quantum advancements. This book attempts to bridge the gap between social intelligence and cybersecurity, offering a holistic and nuanced understanding of these interconnected domains. Through real-world case studies, the author provides practical insights and strategies for adapting to the evolving technological landscape and building a more secure digital future.

This book is intended for futuristic minds, computer engineers, policymakers, or regulatory experts interested in the implications of the revolution of human intelligence on cybersecurity laws and regulations. It will be of interest to cybersecurity professionals and researchers looking for a historic and comprehensive understanding of the evolving landscape, including social intelligence, quantum computing, and algorithm design.

Soorena Merat is the chief cybersecurity officer at Silkatech Consulting Engineers. Dr Merat brings over three decades of unwavering commitment and expertise in the ever-evolving landscape of technology industries. His profound interests lie in the dynamic intersection of artificial intelligence and cybersecurity intelligence testing, explicitly focusing on predicting vulnerabilities and generating actionable control recommendations. His journey in the tech industry has been driven by a passion for innovation, a commitment to cybersecurity and hypersecurity excellence, and a continuous pursuit of knowledge in the rapidly evolving realm of technology.

Cybersecurity 2050
Protecting Humanity in a Hyper-Connected World

Soorena Merat

CRC Press
Taylor & Francis Group
Boca Raton London New York

CRC Press is an imprint of the
Taylor & Francis Group, an **informa** business

First edition published 2026
by CRC Press
2385 NW Executive Center Drive, Suite 320, Boca Raton FL 33431

and by CRC Press
4 Park Square, Milton Park, Abingdon, Oxon, OX14 4RN

CRC Press is an imprint of Taylor & Francis Group, LLC

© 2026 Soorena Merat

ISBN: 978-1-041-07636-0 (hbk)
ISBN: 978-1-041-07635-3 (pbk)
ISBN: 978-1-003-64150-6 (ebk)

DOI: 10.1201/9781003641506

Typeset in Times
by KnowledgeWorks Global Ltd.

Contents

Preface

In the grand narrative of human evolution, the journey from our earliest origins on Earth to the prospect of establishing a foothold on Mars represents a profound odyssey of intelligence, adaptation, and technological prowess. This book embarks on a captivating exploration of this journey, tracing the remarkable trajectory of human intelligence from the primitive habitats of early Earth to the challenges and opportunities presented by the nascent Martian settlements.

Our story begins in the harsh and unforgiving landscapes of early Earth, where survival was paramount, and human intelligence was honed in the crucible of a relentless struggle against the elements. The challenges faced by our ancestors in those primordial environments bear striking similarities to the trials that await the pioneers of Mars. Isolation, resource scarcity, and the ever-present threat of environmental hazards demand resilience, ingenuity, and a profound understanding of the delicate balance between humanity and its surroundings.

However, amidst these challenges, the seeds of human ingenuity were sown. The development of language, art, and social structures fostered cooperation, knowledge sharing, and the transmission of cultural heritage across generations. The emergence of philosophy and scientific inquiry ignited a flame of intellectual curiosity, propelling humanity toward a deeper understanding of the natural world and our place within it.

The invention of tools and technologies extended human capabilities, transforming how we interacted with our environment and each other. From the mastery of fire to the development of agriculture, from the invention of the wheel to the construction of monumental architecture, each technological leap propelled human intelligence forward, reshaping societies and expanding the horizons of possibility.

The rise of the digital age, marked by the invention of electronics and the interconnectedness of the internet, ushered in a new era of human intelligence. The unprecedented access to information, communication, and computational power has transformed how we think, learn, and interact with the world. However, this technological revolution also presents new challenges, particularly in isolated environments like those found on Mars.

As humans venture beyond Earth, establishing outposts on distant planets, they increasingly rely on advanced technologies, including artificial intelligence, for survival. While enabling remarkable feats of exploration and colonization, this dependence also exposes humans to new vulnerabilities, particularly in cybersecurity.

In Mars's isolated and technologically dependent habitats, the threat of cyberattacks looms large. The disruption of critical systems, information manipulation, and trust erosion could have devastating consequences for the fragile Martian settlements. This new frontier demands re-evaluating our understanding of human intelligence, cybersecurity, and the delicate balance between technological dependence and human resilience.

This book delves into these complex issues, exploring the evolution of human intelligence from the early Earth habitats to the challenges and opportunities presented by the Martian frontier. By tracing the trajectory of human ingenuity, adaptation, and technological advancement, we can gain valuable insights into the future of human civilization and the enduring quest to explore, understand, and transcend the boundaries of our world.

Note to readers:

This book is not just a guide but an invitation to a journey. A journey through the labyrinth of technology, where we explore how it shapes our present and how it might sculpt our future, both on this planet and beyond. We invite you, the bright minds and the curious souls, to question, ponder, and engage in a dialogue about the intricate dance between humanity and technology.

This is not a dry technical manual; it is a narrative woven with historical threads, from the invention of the printing press to the rise of social media, revealing how technology has molded communication, social interaction, and even our perception of the world. We will delve into the psychology of cybersecurity, exposing the vulnerabilities of the human mind in the digital age and exploring how our cognitive biases and emotional vulnerabilities can be exploited.

Nevertheless, we go further, venturing beyond today's technology's familiar landscape into tomorrow's uncharted territories. We will explore the challenges and opportunities of artificial intelligence, the ethical dilemmas of a hyper-connected world, and the implications of technology for human settlements on other planets.

While we touch upon technical concepts, our primary focus is igniting awareness and inspiring behavioral change. You will learn to recognize and mitigate cybersecurity risks daily, from identifying phishing scams to protecting your privacy online. However, more importantly, you will be challenged to think critically about the role of technology in your life, your community, and the future of humanity.

Consider this book a stepping stone, a catalyst for deeper exploration. We encourage you to seek insights from cybersecurity experts, stay informed about current events in the field, and engage in discussions about the ethical implications of technology.

Embark on this thought-provoking journey with us, where the past illuminates the present, and proactive awareness becomes your most potent defense. By the end of this book, you will be equipped with the knowledge and tools to navigate the digital world safely and confidently, as well as the critical thinking skills and ethical awareness to shape a future where technology empowers and elevates humanity.

Soorena Merat
Silkatech Consulting Engineers Inc.

1 The Dawn of Cyber Insecurity
Echoes of Early Human Habitats

INTRODUCTION

In the relentless pursuit of technological advancement, we often overlook the echoes of the past that reverberate through our modern world. This chapter embarks on a journey back to the dawn of human civilization, exploring the parallels between the first primitive Earth habitats and cybersecurity challenges in isolated environments. By understanding the security issues early humans face, we can glean valuable insights into the vulnerabilities and resilience of human intelligence in the face of isolation and limited social interaction.

THE FIRST HUMAN HABITATS: ISOLATED AND VULNERABLE

Imagine a world devoid of bustling cities, intricate communication networks, and the constant hum of technology. Early human habitats were starkly different from our modern society, characterized by isolation, scarcity, and the ever-present threat of natural predators. These primitive communities, scattered across vast landscapes, faced challenges that echo the cybersecurity concerns of today's isolated environments, such as space habitats or remote research stations.

Communication in early human habitats was rudimentary, often limited to face-to-face interactions within small, close-knit groups. These early humans relied on vocalizations, gestures, and rudimentary visual signals to convey information. Occasionally, messengers would be dispatched to carry news or warnings across greater distances, but these journeys were perilous and subject to significant delays.

This limited information flow created vulnerabilities that echo those in today's isolated digital environments. Misinformation could quickly spread, fueled by misunderstandings, misinterpretations, or deliberate manipulation. Rumors would swirl, potentially inciting fear, distrust, or even conflict within the group. The lack of reliable communication channels made verifying information or countering false narratives difficult, leaving these early communities susceptible to social disruption.

Interestingly, the early human reliance on animals played a surprising role in shaping these communication patterns. The need to coordinate hunting strategies, share warnings about predators, and track animal migrations pushed humans to develop more sophisticated communication methods. Observations of animal

DOI: 10.1201/9781003641506-1

behavior, such as bird calls or the alarm signals of prey animals, likely inspired early humans to experiment with their vocalizations and signals. Even the domestication of animals, such as dogs and horses, contributed to communication by facilitating the movement of messengers and expanding the range of human interaction.

In essence, the challenges of communicating in a world dominated by the need to interact with animals laid the foundation for developing human language and social structures. The vulnerabilities and adaptations of these early communication systems offer valuable lessons for understanding the importance of secure and reliable information flow in today's interconnected world.

Without established security measures and institutions, early humans depended heavily on trust and reputation within their communities to maintain social order and protect themselves from harm. This reliance on interpersonal relationships, particularly concerning land ownership and access, was crucial in shaping early societal communications. Disputes over land rights and resources often fueled conflicts, highlighting the need for clear communication and reliable information exchange to prevent misunderstandings and maintain peaceful relations. The concurring dependence on land trust and reputation fostered a complex web of alliances and rivalries, shaping the dynamics of early human interactions and laying the foundation for developing more sophisticated communication strategies.

This dependence on interpersonal connections mirrors the importance of trust in modern cybersecurity. Today, social engineering attacks often exploit human vulnerabilities, such as our inclination to trust authority figures or our fear of missing out on gaining access to sensitive information. Just as early humans relied on their social networks to assess the trustworthiness of individuals, modern cybersecurity measures often incorporate reputation-based systems and trust scores to evaluate the credibility of online entities and transactions. Understanding the dynamics of trust and communication in early human societies can provide valuable insights into the psychological underpinnings of social engineering attacks and inform the development of more effective cybersecurity awareness training.

THE RISE OF DECEPTION AND MANIPULATION

Even in primitive societies, where survival was the primary focus, deception and manipulation were tools used to gain an advantage or exploit vulnerabilities. Cunning words, often cloaked in promises of prosperity or safety, could sway individuals and communities toward disastrous decisions. Forged alliances, built on pretenses, could lead to betrayals and conflicts that shattered fragile social structures. Outright trickery, such as disguising intentions or spreading misinformation, could sow discord and distrust among closelyknit groups.

These early forms of manipulation had profound consequences. The erosion of trust within a community hindered cooperation and collective decision-making, making them vulnerable to external threats and internal strife. The fear of deception could lead to isolation and paranoia, further fracturing social bonds.

Though rudimentary compared to today's sophisticated cyberattacks, the manipulative tactics employed by early humans laid the foundation for the social engineering techniques that plague our digital world. The same psychological vulnerabilities that made our ancestors susceptible to cunning words and false promises are exploited by today's cybercriminals, who use phishing scams, fake social media profiles, and other deceptive tactics to access sensitive information and manipulate online behavior.

LESSONS FROM THE PAST: RESILIENCE AND ADAPTATION

Despite the formidable challenges faced by early humans in their isolated habitats, they demonstrated remarkable resilience and adaptability. They developed social norms, communication strategies, and sive mechanisms to mitigate the risks of isolation and deception. These adaptations offer valuable lessons for securing today's isolated environments, underscoring the critical role of human factors in cybersecurity.

By exploring the parallels between early human habitats and modern isolated environments, we gain a deeper appreciation for the enduring challenges of cybersecurity. The vulnerabilities and resilience of human intelligence in the face of isolation and limited social interaction provide crucial insights for developing robust security measures. These measures must focus on technological solutions and consider the psychological and social dynamics influencing human behavior in isolated and interconnected settings.

As we navigate an increasingly complex digital landscape, understanding the human element of cybersecurity becomes ever more critical. By learning from the past and applying those lessons to the present, we can build a more secure and resilient future for individuals and communities. Let us embrace a human-centric approach to cybersecurity that recognizes the enduring power of human ingenuity and adaptability in the face of ever-evolving threats.

THE TAMING OF THE WILD: DOMESTICATION AND THE RISE OF HUMAN INTELLIGENCE

The journey of human evolution is a tapestry woven with ingenuity, adaptation, and transformation threads. Among the most pivotal turning points in this journey was the advent of domestication. This chapter delves into the profound impact of domestication on the development of early human intelligence and the rise of complex societies. By taming wild plants and animals, humans embarked on a path that would forever alter their relationship with the natural world and unleash a cascade of cognitive and social advancements.

DOMESTICATION: A CATALYST FOR COGNITIVE GROWTH

The act of domestication required early humans to engage with the natural world in unprecedented ways, fostering a profound shift in their cognitive abilities. They no longer react to the environment; they begin actively shaping it. For instance, the selective breeding of plants demanded keen observation of traits like size, yield, and disease resistance. This involved noticing differences and understanding the subtle interplay of factors that influenced these traits. Similarly, domesticating animals requires recognizing behavioral patterns, understanding social hierarchies, and applying this knowledge to manage herds effectively.

This intricate dance between observation, understanding, and manipulation spurred the development of critical thinking skills. Early humans had to analyze cause and effect, experiment with different approaches, and adapt their strategies based on the results. The ability to plan and execute long-term strategies, such as saving seeds for future planting or managing breeding cycles, further enhanced their capacity for abstract reasoning and foresight.

These cognitive advancements had far-reaching consequences. Early humans became more adept at problem-solving, innovation, and adapting to changing environments. They developed more sophisticated tools, refined their hunting and gathering techniques, and formed more complex social structures. The foundation was laid for the eventual rise of agriculture, settled communities, and the explosion of technological and cultural innovation that would shape human history.

FROM HUNTER-GATHERERS TO FARMERS AND HERDERS

In the annals of human history, few transitions have been as transformative as the shift from nomadic hunter-gatherer lifestyles to settled agriculture and animal husbandry, and this pivotal moment, often called the Neolithic Revolution, marked a profound departure from the subsistence strategies that had defined human existence for millennia. This chapter delves into the profound impact of this transition, exploring how the domestication of plants and animals laid the foundation for the rise of complex societies and the flourishing of human civilization.

The emergence of agriculture was not a sudden event but a gradual process that unfolded over centuries. Driven by environmental changes, population pressures, and human ingenuity, early humans began cultivating wild plants and experimenting with rudimentary farming techniques. This marked the beginning of a profound shift in human-land relationships, as humans transitioned from passive consumers of nature's bounty to active cultivators of the land.

The domestication of plants and animals provided a stable and predictable food supply, starkly contrasting the uncertainties of the hunter-gatherer lifestyle. This newfound stability had far-reaching consequences. With a surplus of food, populations began to grow, leading to the establishment of permanent settlements and the emergence of villages and towns.

The stability afforded by agriculture also allowed for the specialization of labor. No longer solely focused on securing sustenance, individuals could dedicate their time and energy to developing specific skills and crafts. This led to the emergence of artisans, merchants, priests, and rulers, each contributing to the growing complexity of human societies.

The transition to settled agriculture freed up cognitive resources previously dedicated to the constant pursuit of food. This "cognitive surplus" enabled humans to devote more time and energy to developing new technologies, social structures, and cultural practices. The invention of the wheel, the development of writing, and the construction of monumental architecture are just a few examples of the intellectual flourishing that followed the advent of agriculture.

GATHERING SPARKS: FIRE AND THE FORGING OF EARLY HUMAN SOCIETIES

In the grand narrative of human evolution, the mastery of fire is a pivotal turning point, a technological leap that ignited both flames and the very spark of civilization. This chapter delves into the profound impact of fire on early human societies, exploring how the invention and control of this elemental force transformed our ancestors' lives, fostering the development of micro-societies and laying the foundation for the emergence of societal intelligence.

Fire as a Catalyst: Illuminating the Night, Ward off Predators

The discovery and control of fire marked a watershed moment in human history. No longer at the mercy of the elements, early humans could now illuminate the darkness, ward off predators, and cook food, unlocking a cascade of benefits that would forever alter their social and cognitive landscape.

The ability to generate light and warmth extended the hours of activity beyond daylight, fostering a sense of security and enabling social interaction to continue into the night. The flickering flames of a campfire became a focal point, drawing individuals together and creating a sense of community.

Fire is also protected from predators, offering a powerful deterrent to nocturnal threats and allowing early humans to venture into new territories more confidently. This newfound security fostered a sense of collective strength and encouraged individual cooperation.

THE CULINARY REVOLUTION: COOKING AND COGNITIVE DEVELOPMENT

The invention of cooking, made possible by fire control, revolutionized human nutrition and had profound implications for cognitive development. Cooking not only made food safer and more digestible but also unlocked nutrients and calories that were previously inaccessible.

This dietary shift fueled brain development, expanding cognitive capabilities and the emergence of complex thought processes. Cooking itself, requiring planning, coordination, and the sharing of food, fostered social interaction and the development of communication skills.

MICRO-SOCIETIES: GATHERING AROUND THE HEARTH

The campfire became the heart of early human communities, where individuals shared stories, knowledge, and experiences. These micro-societies huddled around the warmth and light of the flames, fostered a sense of belonging, promoted cooperation, and facilitated the transmission of cultural heritage across generations.

The campfire also served as a focal point for ritual and ceremony, strengthening social bonds and reinforcing shared beliefs. The rhythmic dance of flames, the crackling sounds, and the shared experience of warmth and light created a sense of awe and wonder, contributing to the development of early spiritual and religious practices.

THE BIRTH OF SOCIETAL INTELLIGENCE

The emergence of micro-societies around the campfire laid the foundation for developing societal intelligence. The proximity and frequent interaction among individuals fostered the development of communication skills, social norms, and cooperative behaviors.

The sharing of stories and knowledge around the campfire contributed to transmitting cultural heritage and accumulating collective wisdom. The ability to learn from the experiences of others, to plan and coordinate activities, and to solve problems collectively enhanced the cognitive capabilities of early humans and laid the groundwork for the development of complex societies.

The mastery of fire was a pivotal moment in human history, igniting not only flames but also the spark of civilization. The ability to control this elemental force transformed the lives of early humans, fostering the development of micro-societies and laying the foundation for the emergence of societal intelligence.

The campfire became the heart of these early communities, a gathering place where individuals shared stories, knowledge, and experiences. The warmth and light of the flames fostered a sense of security, promoted cooperation, and facilitated the transmission of cultural heritage.

The invention of cooking, made possible by fire, fueled brain development and contributed to expanding cognitive capabilities. The act of cooking itself fostered social interaction and the development of communication skills.

Micro-societies emerged around the campfire, laying the groundwork for developing societal intelligence. The proximity and frequent interaction among individuals fostered the development of communication skills, social norms, and cooperative behaviors.

The sharing of stories and knowledge around the campfire contributed to transmitting cultural heritage and accumulating collective wisdom. The ability to learn from the experiences of others, to plan and coordinate activities, and to solve problems collectively enhanced the cognitive capabilities of early humans and laid the groundwork for the development of complex societies.

In conclusion, the mastery of fire was a transformative event in human history, igniting civilization's flames and fostering societal intelligence development. The campfire played a crucial role in shaping the course of human evolution and laying the foundation for the complex societies we inhabit today.

THE BIRTH OF CIVILIZATION

The combined effects of a stable food supply, specialized labor, and increased cognitive capacity fueled the rise of civilizations. Cities emerged as centers of trade, innovation, and cultural exchange. Complex social hierarchies, political systems, and religious beliefs developed to manage these growing populations and maintain social order.

The transition from nomadic hunter-gatherers to settled agriculturalists marked a watershed moment in human history. The domestication of plants and animals provided the foundation for the rise of complex societies, the development of new technologies, and the flourishing of human civilization.

This chapter has explored the profound impact of this transition, highlighting the enduring legacy of agriculture in shaping the world we inhabit today.

The surplus of food and resources generated by domestication fueled the rise of complex societies. As humans developed intricate social hierarchies, political systems, and economic networks, villages grew into towns and cities. The need to manage these complex societies further spurred the development of language, writing, and mathematics, accelerating the transmission of knowledge and cultural heritage across generations.

Domestication not only transformed the external world but also shaped the human mind itself. The close interaction with domesticated animals, particularly dogs, may have fostered the development of empathy, cooperation, and social intelligence. The shared experiences of caring for animals and working together toward common goals likely strengthened social bonds and promoted the development of complex communication skills.

Domestication stands as a testament to human ingenuity and adaptability. By taming wild plants and animals, early humans unlocked a cascade of cognitive and social advancements that laid the foundation for modern civilization. The legacy of domestication continues to shape our world today, reminding us of our relationship with the natural world's profound impact on our intellectual and societal development.

THE DARK SHADOW OF SLAVERY: A PARADOXICAL INFLUENCE ON EARLY HUMAN DEVELOPMENT

The narrative of human progress is often interwoven with dark chapters that challenge our understanding of civilization. Slavery, an institution rooted in exploitation and dehumanization, casts a long shadow over our history. This chapter delves into the complex and paradoxical role of slavery in early human societies, exploring its impact on intelligence development and societal advancements while acknowledging its inherent immorality. Slavery, in its essence, represents a gross violation of human rights and dignity. It is a system built on forced labor and the subjugation of individuals, denying them their freedom and autonomy. However, the rise of slavery in early societies also inadvertently fueled certain advancements, albeit at a tremendous moral cost.

FORCED SPECIALIZATION AND TECHNOLOGICAL INNOVATION

The availability of a captive workforce enabled a division of labor that fundamentally reshaped early societies. With a segment of the population forced into manual labor, other individuals were freed from basic survival and subsistence farming demands. This created a social hierarchy where a privileged class could dedicate their time and energy to intellectual or creative pursuits, accelerating the development of new technologies and crafts.

Enslaved people played a crucial role in various industries, contributing significantly to their societies' economic and technological progress. In agriculture, their labor enabled large-scale cultivation and the development of irrigation systems. In mining, they extracted valuable resources that fueled the growth of metallurgy and toolmaking. In construction, their efforts resulted in impressive architectural feats, from pyramids and temples to roads and aqueducts. Furthermore, in the artisanry, their skills produced exquisite works of art, pottery, and textiles. However, it is essential to remember that these advancements were achieved through the brutal exploitation of enslaved people. Their forced labor and dehumanization represent a dark chapter in human history, a stark reminder that progress should never come at the expense of human dignity and freedom.

KNOWLEDGE TRANSFER AND CULTURAL EXCHANGE

Slavery, despite its brutality, also served as a conduit for the transfer of knowledge and cultural practices. As enslaved people were forcibly transported across vast distances, they carried with them a wealth of skills, traditions, and ideas. These skills ranged from agricultural techniques and craftsmanship to medicinal knowledge and artistic expression. Their traditions encompassed languages, religions, music, and culinary practices, enriching the cultural tapestry of the societies they were forced into.

However, this knowledge transfer was not a passive or unidirectional process. Enslaved people actively shaped the cultures of their captors, influencing language, music, cuisine, and even religious beliefs. This complex interplay of cultural exchange, though born out of violence and oppression, highlights the resilience and adaptability of the human spirit.

Despite any advancements fueled by slavery, it is crucial to recognize the deceptive illusion of progress it created. While societies may have benefited economically and technologically from the exploitation of enslaved people, this progress was fundamentally tainted by injustice and suffering. The actual cost of slavery, measured in human lives, shattered families, and the denial of fundamental rights, far outweighs any perceived gains.

The notion that slavery somehow contributed to human progress is a dangerous fallacy. Actual progress cannot be built on a foundation of oppression and dehumanization. The exploitation of enslaved people represents a profound moral failure, and any advancements achieved through their forced labor are a testament to human resilience and ingenuity, not a justification for the institution itself.

THE MORAL RECKONING: CONFRONTING THE LEGACY OF SLAVERY

It is crucial to acknowledge that any advancements attributed to slavery were achieved through the systematic dehumanization and exploitation of individuals. The legacy of slavery continues to haunt modern societies, reminding us of the deep-seated inequalities and injustices that persist.

Slavery inflicted immense suffering, tearing families apart, denying individuals their fundamental human rights, and leaving lasting scars on the social fabric. Its impact reverberates through generations, perpetuating systemic inequalities and hindering the full realization of a just and equitable society. While it is important to recognize any advancements that may have indirectly resulted from slavery, it is equally important to avoid minimizing or justifying the institution's inherent brutality.

The actual cost of slavery cannot be measured solely in economic or technological terms. The psychological trauma, the loss of cultural heritage, and the denial of basic human dignity represent an immeasurable debt that can never be repaid. As we strive for a more just and equitable world, we must remember the lessons of slavery and actively work to dismantle the systemic inequalities that continue to marginalize and oppress vulnerable communities.

The pursuit of progress must always prioritize human dignity and freedom. True advancement lies not in technological or economic gains alone but in creating a society where everyone can reach their full potential and live free from exploitation and oppression.

THE RISE OF ARTISTIC EXPRESSION: A TAPESTRY OF HUMAN INTELLIGENCE AND SOCIETAL ADVANCEMENTS

The human narrative is interwoven with threads of creativity, innovation, and the relentless pursuit of self-expression. From the earliest cave paintings to the grand symphonies of the modern era, art has been an inseparable companion on this journey. This chapter explores the profound impact of art on the development of early human intelligence and the advancement of societies. By delving into the creative expressions of our ancestors, we can unravel the intricate tapestry of human ingenuity and societal evolution.

ART AS A CRUCIBLE OF COGNITIVE DEVELOPMENT

In its myriad forms, art served as a crucible for developing early human intelligence. The act of creating art demanded observation, imagination, and problem-solving skills. Early humans had to experiment with various materials, tools, and techniques to translate their creative visions into tangible forms. This process honed their cognitive abilities, fostering critical thinking, spatial reasoning, and abstract thought.

THE BIRTH OF SYMBOLIC THINKING

The emergence of art marked a pivotal moment in human cognitive evolution: the birth of symbolic thinking. Cave paintings, sculptures, and early decorative artifacts demonstrate the ability of early humans to represent abstract concepts, ideas, and emotions through visual and tangible mediums. This capacity for symbolic representation laid the foundation for developing language, writing, and complex communication systems.

ART AS SOCIAL GLUE

Art was crucial in strengthening social bonds and fostering community among early humans. Communal artistic endeavors, such as creating cave paintings or performing rituals and dances, fostered cooperation, shared experiences, and a sense of belonging. Art was a powerful tool for transmitting cultural knowledge, values, and traditions across generations, contributing to the continuity and stability of early societies.

The evolution of art reflects the advancements of early human societies. As communities grew and technologies developed, artistic expressions became more sophisticated and diverse. The invention of new tools and materials, such as pigments, pottery, and musical instruments, expanded the creative possibilities for early artists. The emergence of social hierarchies and organized religions also influenced artistic themes and styles, reflecting the changing social and cultural landscape.

Art reflected societal advancements and served as a catalyst for innovation. The creative exploration of new artistic techniques and materials often led to the development of new technologies and applications that extended beyond art. For example, the development of pottery-making techniques contributed to advancements in metallurgy and the creation of new tools and weapons.

Art has been integral to the human journey since its earliest beginnings. It has served as a crucible for cognitive development, a catalyst for innovation, and a powerful tool for social cohesion and cultural transmission. This chapter explores the profound impact of art on the development of early human intelligence and the advancement of societies, highlighting the enduring legacy of artistic expression in shaping the world we inhabit today.

While often viewed through a modern lens of aesthetics and individual expression, art played a fundamentally different role in early human societies. It was the lifeblood of the community, a powerful force shaping social bonds, transmitting knowledge, and ensuring cultural continuity. This chapter explores the vital role of art in knitting together the fabric of early human groups, from fostering cooperation and shared identity to preserving and passing on traditions in a world without written language.

Imagine a group of early humans gathered around a fire, their voices raised in a rhythmic chant as they dance and paint on the cave walls. This collective artistic experience transcended mere decoration; it was an act of community building. The shared effort, the synchronized movements, and the collective focus on a single creative goal fostered a profound sense of unity and belonging.

Cave paintings, depicting hunts, animals, and mythical beings, served as visual narratives, communicating shared experiences and beliefs. These early forms of storytelling strengthened social bonds by creating a shared understanding of the world and the group's place within it. Through art, early humans could pass on knowledge about successful hunting strategies, warn of potential dangers, and celebrate their triumphs, fostering a sense of shared history and purpose.

Ritualistic dances and musical performances, often incorporating elaborate costumes and body art, were not merely entertainment. They were powerful social rituals that reinforced group identity and solidified social structures. The synchronized movements and rhythmic patterns created a sense of collective harmony, while the elaborate preparations and performances instilled a sense of shared purpose and tradition. Without written language, art was the primary vehicle for transmitting cultural knowledge, values, and traditions across generations. Through intricate carvings, symbolic ornaments, and oral traditions interwoven with music and dance, early humans passed on their accumulated wisdom, ensuring the survival and continuity of their communities. Each generation added its layer of creativity and innovation, enriching the cultural tapestry and strengthening the bonds that held their society together. Art in early human societies was more than just aesthetics; it was the community's beating heart. It fostered cooperation, forged a shared identity, and preserved cultural heritage. By understanding the vital role of art in these early groups, we gain a deeper appreciation for its power to shape human connections and drive societal development.

ART AS A MIRROR OF SOCIETAL ADVANCEMENTS

Throughout history, art has served as a powerful lens through which we can glimpse the intricate tapestry of human civilization. More than just aesthetic creations, artistic expressions encapsulate societies' values, beliefs, and aspirations, providing invaluable insights into their evolution. This chapter explores the dynamic interplay between art and the advancement of early human societies, demonstrating how artistic creations mirrored and influenced our ancestors' social, cultural, and technological transformations. The evolution of art provides a fascinating chronicle of the progress of early human societies. From the rudimentary cave paintings of the Paleolithic era to the intricate pottery and sculptures of the Neolithic period, artistic creations serve as tangible markers of societal development. As communities grew and technologies advanced, artistic expressions mirrored these changes, becoming more sophisticated and diverse. The invention of new tools and materials played a crucial role in expanding the creative possibilities for early artists. The discovery of pigments allowed for the vibrant expression of colors in cave paintings and rock art, while the development of pottery enabled the creation of functional and decorative vessels. The invention of musical instruments, such as flutes and drums, added a new dimension to artistic expression, enriching cultural ceremonies and social gatherings.

The emergence of social hierarchies and organized religions profoundly influenced artistic themes and styles. As societies became more complex, art began to reflect the time's power dynamics and social structures. The construction of monumental architecture, such as pyramids and temples, showcased the authority of rulers and religious leaders, while intricate carvings and sculptures commemorated their achievements and reinforced their status.

Art also served as a powerful medium for expressing and transmitting cultural values and beliefs. Early cave paintings' depictions of animals, hunting scenes, and fertility symbols reflect the close relationship between humans and the natural world. The emergence of religious iconography and symbolic representations in later periods highlights the growing importance of spirituality and belief systems in shaping social identity and cultural practices.

Art has been an inseparable companion on the journey of human civilization, reflecting and influencing the social, cultural, and technological advancements of societies. This chapter has explored the dynamic interplay between art and the evolution of early human societies, demonstrating how artistic creations served as a mirror, capturing the essence of each era and providing invaluable insights into the human story. By understanding the role of art in shaping our past, we can gain a deeper appreciation for its enduring power to inspire, challenge, and connect us as a global community.

Art reflected societal advancements and served as a catalyst for innovation. The creative exploration of new artistic techniques and materials often led to the development of new technologies and applications that extended beyond art. For example, the development of pottery-making techniques contributed to advancements in metallurgy and the creation of new tools and weapons. Art has been integral to the human journey since its earliest beginnings. It has served as a crucible for cognitive development, a catalyst for innovation, and a powerful tool for social cohesion and cultural transmission. This chapter explores the profound impact of art on the development of early human intelligence and the advancement of societies, highlighting the enduring legacy of artistic expression in shaping the world we inhabit today.

THE DAWN OF WORDS: LANGUAGE AND THE ASCENT OF HUMAN INTELLIGENCE

In the grand tapestry of human evolution, the emergence of language stands as a defining moment, a spark that ignited the flames of complex thought, social cooperation, and cultural transmission. This chapter delves into the profound impact of language on the development of early human intelligence and the advancement of societies. By tracing the origins and evolution of language, we can unravel the intricate web of cognitive and social transformations that propelled humanity toward its unique position in the world. Language, with its intricate structure and symbolic representations, served as a crucible for the development of early human intelligence. Language acquisition and use require the development of sophisticated cognitive abilities, including memory, attention, and abstract reasoning. The ability to categorize objects, understand relationships, and express complex ideas through words fostered the growth of critical thinking, problem-solving, and creative expression.

FROM SOUNDS TO SYMBOLS: THE BIRTH OF COMMUNICATION

The capacity for language, a hallmark of human intelligence, sets us apart from all other species on Earth. It is a symphony of sounds, symbols, and syntax that allows us to communicate complex ideas, share knowledge, and build intricate social structures. This chapter embarks on a journey through the evolutionary odyssey of language, tracing its origins from rudimentary vocalizations to the sophisticated communication systems we use today.

THE SEEDS OF LANGUAGE: PRECURSORS TO COMPLEX COMMUNICATION

The roots of language can be traced back to the pre-linguistic communication of our primate ancestors. These early forms of communication relied heavily on gestures, facial expressions, and vocalizations to convey basic needs and emotions. While lacking the complexity of human language, these rudimentary systems laid the foundation for developing more sophisticated communication.

VOCALIZATION: THE DAWN OF SPOKEN LANGUAGE

The transition from simple vocalizations to complex speech was a gradual process shaped by biological and social factors. Early humans likely began with a limited repertoire of sounds, gradually expanding their vocal range and combining sounds to create new meanings. This process was likely driven by the need for more precise communication as social groups grew and interactions became more complex.

THE EMERGENCE OF STRUCTURE: PROTO-LANGUAGES AND BEYOND

As vocalizations became more nuanced, proto-languages emerged, characterized by rudimentary grammatical structures and a growing vocabulary. These early languages likely relied on sounds, gestures, and contextual cues to convey meaning. Over time, these proto-languages evolved, incorporating more complex grammatical rules and expanding their lexicon to encompass a broader range of concepts and ideas.

THE SYMBOLIC LEAP: FROM SOUNDS TO MEANING

A pivotal moment in the evolution of language was the development of symbolic representation. This cognitive leap allowed humans to associate specific sounds with abstract concepts, objects, and actions. Using symbols to represent the world around them unlocked a new level of communicative power, enabling humans to express complex thoughts, share knowledge, and build intricate social structures.

THE BIRTH OF GRAMMAR: SHAPING THE STRUCTURE OF LANGUAGE

The emergence of grammar marked a significant milestone in the development of language. Grammatical rules provided a framework for organizing words and phrases, creating complex sentences, and expressing nuanced meanings. The development of grammar was likely driven by the need for greater clarity and precision in communication, mainly as human societies grew and interactions became more complex.

LANGUAGE AS A CULTURAL TOOL

Language facilitated communication and served as a powerful tool for cultural transmission. Through stories, songs, and rituals, early humans passed down their knowledge, beliefs, and traditions from one generation to the next. Language became a vehicle for preserving cultural heritage, shaping social identity, and transmitting the accumulated wisdom of human experience. The journey from simple vocalizations to complex language systems was a remarkable feat of human ingenuity and adaptation. This chapter has explored the key milestones in this evolutionary odyssey, highlighting the gradual emergence of structure, the symbolic leap, and the development of grammar. By understanding the origins and evolution of language, we gain a deeper appreciation for its profound impact on human intelligence, social interaction, and cultural development. Language is a testament to our unique capacity for communication, creativity, and the transmission of knowledge, shaping the world we inhabit today and paving the way for future generations to build upon the rich legacy of human expression.

The language was pivotal in strengthening social bonds and fostering cooperation among early humans. Communicating complex ideas and sharing knowledge facilitated the coordination of activities, the transmission of cultural traditions, and the development of social norms. Language enabled humans to form intricate social networks, resolve conflicts, and build cohesive communities, laying the foundation for today's complex societies. Language facilitated social interaction and served as a catalyst for innovation and technological advancement. The ability to share knowledge, discuss ideas, and plan strategies allowed early humans to collaborate on complex tasks, develop new technologies, and adapt to changing environments. Language fueled the transmission of knowledge across generations, accelerating the accumulation of wisdom and the advancement of human civilization.

THE EVOLUTION OF THOUGHT

Language profoundly influences the way humans think and perceive the world. The ability to categorize objects, express abstract concepts, and construct narratives shaped our understanding of reality and our place within it. Language enabled humans to reflect on the past, contemplate the future, and engage in complex reasoning and problem-solving, paving the way for developing philosophy, science, and art. Language is a testament to humans' remarkable cognitive and social capabilities. Its emergence marked a turning point in our evolutionary journey, unleashing a cascade of intellectual and societal advancements. This chapter has explored the profound impact of language on the development of early human intelligence and the rise of complex societies, highlighting its enduring legacy in shaping the world we inhabit today. As we continue to explore the complexities of language and its role in human cognition, we gain a deeper appreciation for its power to connect, inspire, and transform.

In the tapestry of early human societies, language served as a thread that connected individuals, communities, and cultures. However, the impact of language extended beyond simple communication; learning other languages played a crucial role in expanding societal intelligence and driving advancements on multiple fronts.

Learning other languages enhanced cognitive flexibility and problem-solving skills. The mental agility required to switch between different linguistic systems fostered a more adaptable and creative mind, better equipped to navigate the complexities of a changing world. This cognitive flexibility likely contributed to innovation and problem-solving in various domains, from tool-making and hunting strategies to social organization and conflict resolution.

Multilingualism also facilitated trade, alliances, and cultural exchange between groups. The ability to communicate across language barriers fostered understanding and cooperation, enabling the exchange of goods, ideas, and technologies. This cross-cultural pollination enriched societies, leading to the diffusion of knowledge and the adoption of beneficial practices from other communities.

Furthermore, the ability to communicate in multiple languages played a crucial role in conflict resolution and diplomacy. By understanding the perspectives and intentions of other groups, early humans could potentially mitigate misunderstandings, negotiate agreements, and build alliances, fostering peaceful coexistence and cooperation.

In conclusion, learning other languages in early human societies was more than just a linguistic skill; it was a catalyst for cognitive enhancement, social cohesion, and cultural exchange. By fostering understanding and cooperation across language barriers, multilingualism plays a pivotal role in shaping our interconnected world today, highlighting the enduring power of language to connect, inspire, and transform.

THE TAPESTRY OF FAITH: RELIGION'S ROLE IN EARLY HUMAN INTELLIGENCE AND SOCIETAL DEVELOPMENT

The human narrative is interwoven with a relentless quest for meaning, understanding, and connection with something greater than oneself. In its myriad forms, religion has been an omnipresent companion on this journey, shaping our beliefs, values, and social structures. This chapter delves into the profound impact of religion on the development of early human intelligence and the advancement of societies. By exploring religious beliefs and practices' historical and cultural significance, we can unravel the intricate tapestry of human ingenuity and societal evolution.

RELIGION AS A CRUCIBLE OF COGNITIVE DEVELOPMENT

From the dawn of consciousness, humanity has gazed at the stars, the storms, and the cycle of life and death with a sense of wonder and trepidation. The need to understand these mysteries, to find meaning and order in the apparent chaos of existence, gave birth to religion. This chapter explores how the development of religious beliefs and practices, with their intricate narratives and rituals, acted as a powerful catalyst in the evolution of early human intelligence.

Religion, in its essence, is an attempt to grapple with the profound questions of existence. Why are we here? What happens after death? What unseen forces govern the world around us? The very act of wrestling with these existential questions stimulated cognitive growth in early humans. Formulating explanations for the unknown and weaving narratives that provided solace and understanding required a burgeoning capacity for abstract thought and complex reasoning.

The development of religious beliefs demanded a cognitive leap beyond the concrete world of immediate perception. Early humans began conceptualizing unseen forces and entities, attributing agency and intention to the natural world. This ability to think abstractly, to imagine beings and powers beyond the realm of the senses, marked a significant advancement in human intelligence. It laid the groundwork for developing symbolic thinking, a cornerstone of human cognition that enabled the creation of language, art, and complex social structures.

Religious practices, with their intricate rituals and ceremonies, further fueled the development of cognitive skills. The performance of rituals required memorization, sequencing, and the ability to follow complex symbolic instructions. These practices engaged multiple cognitive processes, strengthening memory, attention, and problem-solving abilities. Moreover, the shared experience of rituals fostered social cohesion and cooperation, further enhancing social intelligence and group dynamics.

The development of religious narratives, with their stories of creation, gods, and heroes, stimulated the imagination and fostered creative thinking. The need to explain natural phenomena, to understand the human place in the cosmos, and to cope with the uncertainties of life and death drove early humans to develop elaborate mythologies and belief systems. This process nurtured the capacity for storytelling, a fundamental human skill that entertained and transmitted knowledge, values, and cultural traditions across generations.

Religion, with its intricate tapestry of beliefs, rituals, and narratives, played a pivotal role in shaping the cognitive landscape of early humans. The quest for meaning, the grappling with existential questions, and the development of symbolic thinking fostered the growth of imagination, critical thinking, and problem-solving skills. As we explore the historical and cultural significance of religion, we gain a deeper appreciation for its profound impact on the development of human intelligence and the advancement of societies.

THE BIRTH OF SYMBOLIC THINKING AND RITUALISTIC PRACTICES

With its insatiable curiosity and yearning for meaning, the human mind has always sought to understand the world around it. In the dawn of our species, when the forces of nature seemed awe-inspiring and terrifying, religion emerged to explain the inexplicable, find order in chaos, and connect with something larger than oneself. This chapter explores the profound impact of religion on the cognitive evolution of early humans, highlighting the emergence of symbolic thinking and ritualistic practices as key milestones in the development of human intelligence and social organization.

The emergence of religion marked a watershed moment in human cognitive development, giving rise to the capacity for symbolic thinking. Early humans, confronted with the natural world's mysteries, began to associate natural phenomena, objects, and animals with supernatural powers and divine beings. The sun, moon, and stars became celestial deities, while animals were imbued with spiritual significance and revered as totems or spirit guides. This ability to imbue the physical world with symbolic meaning represented a profound cognitive leap, laying the foundation for developing abstract thought and complex communication systems.

Ritualistic practices, often performed in groups, were crucial in reinforcing shared beliefs and fostering social cohesion. Early humans gathered around fires, chanting incantations, performing dances, and enacting symbolic rituals to appease the gods, ensure successful hunts, or commemorate important events. These shared experiences created a sense of collective identity, strengthened social bonds, and instilled a sense of order and predictability in a chaotic and unpredictable world.

The symbolic representations embedded in religious beliefs and practices paved the way for language development. As early humans sought to communicate their experiences of the sacred, they developed a vocabulary of symbols, gestures, and vocalizations to express their beliefs and share their stories. These early forms of communication gradually evolved into more complex language systems, enabling humans to convey abstract ideas, transmit knowledge across generations, and build intricate social structures.

The impulse to express religious beliefs and experiences also found an outlet in artistic creations. Cave paintings, sculptures, and early decorative artifacts often depicted religious themes, showcasing reverence for nature, the worship of deities, and performing rituals. Art was a visual language that communicated shared beliefs and reinforced cultural identity within early human communities.

The emergence of religion marked a profound shift in human cognitive and social evolution. The development of symbolic thinking, ritualistic practices, and complex communication systems laid the foundation for the advancements that would shape human civilization. This chapter explores the intricate interplay between religion, cognition, and social organization, highlighting the enduring legacy of religious beliefs and practices in shaping the human story. By understanding the role of religion in our past, we can gain a deeper appreciation for its profound impact on human intelligence, social dynamics, and cultural evolution.

RELIGION AS A SOCIAL GLUE

Religion acted as a potent unifying force in early human societies, weaving the threads of individual beliefs and communal practices into a cohesive social fabric. Shared beliefs and rituals, often centered around the reverence for deities or ancestral spirits, fostered a sense of belonging and collective identity. These shared experiences, whether through communal prayers, sacred ceremonies, or the observance of religious festivals, promoted cooperation and mutual support within the group. Moreover, religion provided a framework for moral and ethical behavior, establishing guidelines for conduct and reinforcing social norms. Religious institutions, often led by shamans, priests, or elders, were vital social and cultural life hubs. These institutions provided spaces for worship and spiritual guidance and acted as platforms for education, conflict resolution, and the transmission of cultural knowledge and traditions across generations. Religion was a cornerstone of early human societies, fostering unity, promoting cooperation, and shaping the moral and cultural landscape.

RELIGION AS A CATALYST FOR SOCIETAL ADVANCEMENTS

While often viewed solely through the lens of spirituality and faith, religion has played a multifaceted role in advancing human civilization. This chapter explores the often-overlooked ways religion catalyzed societal progress, driving intellectual inquiry, fostering social cohesion, and inspiring remarkable feats of engineering and artistic expression.

The inherent human desire to comprehend the cosmos and our place within it found a powerful outlet in religious expression. Early civilizations often intertwined their cosmological beliefs with the study of astronomy, meticulously tracking celestial movements and developing sophisticated calendars to predict seasonal changes and agricultural cycles. The construction of monumental structures like Stonehenge and the Egyptian pyramids, often aligned with astronomical events, demonstrates the intertwining of religious beliefs and scientific observation.

Furthermore, pursuing spiritual understanding led to the development of philosophical and theological systems grappling with questions of existence, morality, and the afterlife. These intellectual explorations laid the groundwork for developing logic, ethics, and metaphysics, shaping the course of human thought for millennia. Religious institutions often served as vital centers of learning and cultural development. Monasteries and temples housed libraries and scriptoria, preserving and disseminating knowledge through carefully copying manuscripts. Religious patronage supported the creation of breathtaking works of art, from the intricate mosaics of Byzantine churches to the soaring cathedrals of medieval Europe. Music flourished under the auspices of religion, with sacred chants and hymns evolving into complex polyphonic compositions. The pursuit of scientific inquiry also found support within religious institutions. While often portrayed as being in conflict, religion and science have a long history of intertwined development. Many early scientists and philosophers were also religious figures, and their investigations into the natural world were often motivated by a desire to understand the divine order of creation.

Religion played a pivotal role in shaping social structures and establishing moral frameworks. Shared beliefs and rituals fostered a sense of community and belonging, promoting cooperation and social cohesion. Religious narratives and moral codes guided ethical behavior, shaping legal systems and influencing social norms. The concept of divine justice and the promise of reward or punishment in the afterlife were powerful motivators for adherence to social norms and ethical conduct.

Religion's influence extends far beyond the realm of spirituality. This chapter has explored how religion served as a catalyst for societal advancements, driving intellectual inquiry, fostering social cohesion, and inspiring remarkable feats of engineering and artistic expression. By recognizing the complex interplay between religion and human progress, we gain a deeper appreciation for the diverse forces that have shaped the course of human civilization.

THE EVOLUTION OF RELIGIOUS BELIEFS

The evolution of religious beliefs reflects early human societies' changing social and cultural landscape; as communities grew and interacted with other cultures, religious beliefs adapted and diversified. The emergence of organized religions, with their complex theologies, rituals, and social hierarchies, further influenced the development of human intelligence and societal structures.

Religion has been an inseparable companion on the journey of human civilization, shaping our beliefs, values, and social structures. This chapter has explored the profound impact of religion on the development of early human intelligence and the advancement of societies, highlighting its enduring legacy in shaping the world we inhabit today. By understanding the role of religion in our past, we can gain a deeper appreciation for its complex interplay with human cognition, social dynamics, and cultural evolution.

AWAKENING THE MIND: PHILOSOPHY'S ENDURING LEGACY IN EARLY HUMAN DEVELOPMENT

The human journey is an epic odyssey driven by an insatiable curiosity, a relentless yearning to decipher the enigmas of existence and unravel the intricate tapestry of the cosmos. From the dawn of consciousness, our ancestors gazed upon the celestial ballet of stars, pondered the mysteries of life and death, and sought answers to the profound questions that echoed through their minds. This innate thirst for knowledge, this burning desire to comprehend the world and our place within it, gave rise to philosophy, the "love of wisdom."

Emerging from the mists of time, philosophy served as a beacon illuminating the path of human understanding. It challenged the boundaries of conventional thought, encouraging early humans to question, analyze, and seek truth beyond the confines of myth and tradition. This intellectual awakening, fueled by the power of reason and critical inquiry, laid the cornerstone for developing human intelligence and advancing societies.

This chapter embarks on a captivating exploration of philosophy's profound impact on the human journey. We will trace the origins of philosophical thought, delving into the musings of ancient thinkers who dared to challenge the status quo and expand the frontiers of knowledge. We will witness how their ideas, like seeds sown in fertile ground, blossomed into the foundational principles of logic, ethics, and metaphysics, shaping the course of human thought for millennia.

From the shores of ancient Greece to the bustling intellectual centers of the East, we will encounter the great minds that shaped the philosophical landscape. We will delve into the dialogues of Socrates, who challenged conventional wisdom with his relentless questioning. We will explore the profound insights of Plato, who illuminated the nature of reality and the pursuit of the ideal state. We will also grapple with Aristotle's comprehensive system, whose inquiries span the breadth of human knowledge, from logic and ethics to physics and metaphysics.

Through their intellectual endeavors, these philosophical pioneers expanded the boundaries of human understanding and laid the groundwork for societal progress.

Their ideas challenged traditional hierarchies, fostered a spirit of critical inquiry, and inspired generations to seek truth and knowledge through reason and observation. Their legacy continues to resonate today, reminding us of the enduring power of philosophy to shape our understanding of the world, guide our ethical choices, and inspire us to build a more just and enlightened society.

THE BIRTH OF REASON: EARLY PHILOSOPHICAL INQUIRY

As humanity emerged from the shadows of myth and superstition in the dawn of civilization, a new light began to shine – the light of reason. Amidst the ancient world's bustling city-states and nascent empires, a profound shift occurred in the human mind. No longer content with relying solely on mythological explanations for the universe's workings, early thinkers embarked on a quest for knowledge grounded in observation, logic, and critical inquiry. This marked the birth of philosophy, the "love of wisdom," a pursuit that would forever alter the trajectory of human civilization.

The fertile intellectual landscape of ancient Greece gave rise to the pre-Socratic philosophers, pioneers who dared to question the traditional narratives and seek rational explanations for natural phenomena, human behavior, and the very nature of reality. Thales of Miletus, often hailed as the "father of Western philosophy," challenged the prevailing mythological accounts of creation by proposing that water was the arche, the fundamental principle from which all else is derived. This marked a radical departure from supernatural explanations, paving the way for a more naturalistic and scientific worldview.

Anaximander, another Milesian philosopher, further expanded the boundaries of human understanding by positing the Apeiron, an undefined, boundless primordial substance, as the origin of all things. This concept challenged the limitations of human perception and introduced the notion of an underlying principle that transcends the tangible world.

Heraclitus of Ephesus, known for his enigmatic pronouncements and emphasis on the ever-changing nature of reality, offered a dynamic perspective on the cosmos. His famous dictum, "No man ever steps in the same river twice," encapsulates his doctrine of flux, highlighting the constant transformation that characterizes the universe and human experience.

These early philosophical inquiries, though seemingly abstract, laid the foundation for the development of critical thinking, scientific inquiry, and the pursuit of knowledge for its own sake. With their bold questioning and relentless pursuit of truth, the pre-Socratic philosophers ignited a flame of intellectual curiosity that would illuminate the path of human progress for millennia to come.

SOCRATES, PLATO, AND ARISTOTLE:
PILLARS OF WESTERN THOUGHT

In the heart of ancient Greece, amidst the bustling city-states and vibrant cultural landscape, a golden age dawned in Athens, a period of unprecedented intellectual and artistic flourishing. From this fertile ground emerged three towering figures who would forever alter the course of Western philosophy: Socrates, Plato, and Aristotle.

Socrates, a gadfly of the Athenian marketplace, challenged conventional wisdom with his relentless questioning and unwavering pursuit of truth. His method, known as the Socratic method, employed a dialectic approach, engaging individuals in probing conversations that exposed contradictions and inconsistencies in their beliefs. This emphasis on critical self-examination laid the groundwork for the development of rational inquiry and the pursuit of knowledge through reasoned argumentation.

Socrates' most famous student, Plato, inherited his mentor's passion for truth but channeled it into a more systematic and comprehensive philosophical system. In his renowned dialogues, Plato explored the nature of knowledge, justice, and the ideal state, weaving intricate philosophical arguments with literary artistry. His concept of the Forms, eternal and unchanging ideals beyond sensory experience, profoundly influenced Western metaphysics and epistemology.

Aristotle, Plato's student and a polymath of extraordinary breadth, further expanded the horizons of philosophical inquiry. With a mind that embraced empirical observation and theoretical abstraction, Aristotle systematized knowledge across various disciplines, from logic and ethics to physics and metaphysics. His meticulous categorization of knowledge and his emphasis on empirical evidence laid the foundation for the scientific method, shaping the course of scientific inquiry for centuries to come.

These three philosophical giants, each with their unique approach and contributions, laid the cornerstone for Western intellectual tradition. Their ideas reverberated through the ages, influencing philosophers, theologians, scientists, and political thinkers across cultures and continents. Their legacy inspires and challenges us today, reminding us of the enduring power of reason, critical inquiry, and the pursuit of knowledge in shaping the human journey.

THE EXPANSION OF HUMAN UNDERSTANDING

The contributions of these early philosophers extended far beyond abstract theorizing. Their inquiries into the nature of logic and reasoning laid the foundation for developing mathematics, science, and law. Their explorations of ethics and morality shaped social norms and political systems. Their pursuit of knowledge for its own sake fostered a culture of intellectual curiosity and critical thinking that propelled human societies toward greater understanding and progress.

The impact of philosophy on societal advancement is undeniable. The ideas of these early thinkers challenged traditional hierarchies, questioned the status quo, and encouraged individuals to seek truth and knowledge through reason and observation. This intellectual ferment fostered a spirit of innovation and critical inquiry that fueled advancements in science, technology, and social organization.

Philosophy, the "love of wisdom," has been an indispensable companion on the human journey. Its emergence marked a turning point in our intellectual development, fostering critical thinking, expanding our understanding of the world, and laying the foundation for societal progress. This chapter has explored the profound impact of philosophical inquiry, from the musings of ancient thinkers to its enduring legacy in shaping the foundations of human thought and civilization. As we grapple with the fundamental questions of existence, the spirit of philosophical inquiry remains a vital force in our ongoing quest for knowledge, meaning, and a more just and enlightened world.

FROM CANVAS TO CLOCKWORK: HOW PHILOSOPHY, ART, AND LANGUAGE FORGED THE PATH TO MODERNITY

The rise of modern society, with its technological marvels and intricate clockwork mechanisms, was not a sudden eruption but rather the culmination of a long and winding journey. This chapter explores the pivotal role played by advancements in philosophy, art, and language in early societies, laying the groundwork for the scientific revolution and the technological innovations that define our modern world.

Early philosophical inquiry, emphasizing reason, logic, and observation, laid the foundation for a new world understanding. In their quest to identify the fundamental principles of the cosmos, the pre-Socratic philosophers challenged traditional mythological explanations and paved the way for scientific investigation. The development of formal logic, notably by Aristotle, provided a framework for rigorous reasoning and critical thinking, essential tools for scientific exploration.

The artistic expressions of early humans, far from mere decoration, played a crucial role in honing observational skills and fostering creativity. Translating the natural world onto cave walls or shaping clay into intricate forms required careful observation, spatial reasoning, and an imaginative leap. These skills, honed through artistic practice, proved invaluable in developing technologies and engineering feats that would later transform society.

The emergence of language, with its capacity for abstract thought and complex communication, was a watershed moment in human history. Language enables sharing knowledge, transmitting ideas across generations, and the collaborative pursuit of innovation. The development of writing systems further amplified the power of language, allowing for the preservation and dissemination of knowledge on an unprecedented scale.

THE CONVERGENCE OF IDEAS: A FERTILE GROUND FOR INNOVATION

The advancements in philosophy, art, and language converged to create a fertile ground for innovation. The spirit of inquiry fostered by philosophy, the observational skills honed through art, and the communicative power of language combined to propel early societies toward discoveries and technological breakthroughs.

FROM PHILOSOPHY TO SCIENCE: THE BIRTH OF THE SCIENTIFIC METHOD

The burgeoning emphasis on reason and empirical observation, first cultivated in the philosophical inquiries of antiquity, gradually blossomed into the scientific method, a systematic and rigorous approach to unravelling the mysteries of the natural world. This paradigm shift, driven by a thirst for knowledge that transcended the limitations of dogma and speculation, championed observation, experimentation, and meticulous analysis as the cornerstones of understanding. The contributions of towering figures like Copernicus, who dared to challenge the geocentric view of the universe, Galileo, whose telescopic observations unveiled the celestial dance of planets and moons, and Newton, whose laws of motion and gravity illuminated the underlying order of the cosmos, revolutionized our comprehension of the universe and laid the very foundation for the technological advancements that would shape the modern world. This shift toward empirical inquiry and rational explanation ignited a flame of discovery that continues illuminating the path of human progress.

THE RISE OF MACHINES: FROM CLOCKWORK TO THE INDUSTRIAL REVOLUTION

The development of intricate clockwork mechanisms in the medieval era showcased the growing sophistication of human ingenuity. These intricate devices, with their precisely engineered gears and springs, not only served as timekeeping tools but also inspired further innovation in mechanical engineering. The principles of clockwork design found application in various fields, from navigation and astronomy to textile production and weaponry, paving the way for the Industrial Revolution and the mass production of goods.

The advancements in philosophy, art, and language in early societies were not isolated developments but interconnected threads that wove the tapestry of human progress. The spirit of inquiry, the cultivation of observation and imagination, and the power of communication converged to create a fertile ground for innovation, ultimately leading to the rise of modern society with its technological marvels and intricate clockwork mechanisms. This chapter has explored the pivotal role of these early advancements, highlighting their enduring legacy in shaping today's world.

The emergence of mechanical marvels marked a turning point in human history, revolutionizing technology and profoundly impacting human intelligence and societal development. These intricate creations, born from the convergence of philosophical inquiry, artistic expression, and linguistic prowess, extended human capabilities and reshaped the very fabric of society.

The invention of the printing press, a mechanical marvel in its own right, democratized knowledge and fueled the Renaissance. The ability to mass-produce books and disseminate information on an unprecedented scale broke the monopoly of knowledge held by the elite, empowering individuals and fostering a culture of intellectual curiosity. This newfound access to information stimulated critical thinking, broadened perspectives, and fueled a spirit of innovation across all strata of society.

The development of clocks and other timekeeping devices revolutionized daily life and transformed human perception of time. The ability to measure and quantify time with increasing precision fostered a sense of order and predictability, enabling the coordination of complex activities and the efficient organization of labor. This newfound mastery of time had profound implications for productivity, economic growth, and the development of industrial societies.

The invention of navigational instruments, such as the astrolabe and the compass, expanded human horizons and fueled the Age of Exploration. These mechanical marvels enabled sailors to navigate the vast oceans, leading to the discovery of new lands, the exchange of cultures, and the expansion of trade networks. The ability to explore and map the world fostered a sense of interconnectedness and broadened human understanding of the planet's diverse geography and cultures.

The rise of mechanical marvels transformed technology and reshaped human cognition. The intricate workings of machines challenged individuals to think in new ways, fostering problem-solving skills, spatial reasoning, and an understanding of complex systems. The ability to design, build, and operate machines required combining theoretical knowledge and practical skills, further bridging the gap between intellectual inquiry and real-world applications.

The impact of mechanical marvels extended far beyond technology. These innovations transformed social structures, economic systems, and cultural practices. The mass production of goods fueled economic growth and led to the rise of industrial cities. The increased efficiency of transportation and communication networks connected communities and facilitated the exchange of ideas and goods on a global scale.

The rise of mechanical marvels marked a pivotal moment in human history, revolutionizing technology, human intelligence, and societal development. These intricate creations, born from the convergence of philosophical inquiry, artistic expression, and linguistic prowess, extended human capabilities, reshaped our understanding of the world, and laid the foundation for our modern societies.

The emergence of mechanical marvels sparked a revolution in human intelligence, propelling societies toward new heights of knowledge, innovation, and interconnectedness. These inventions extended human capabilities and transformed how we communicate, learn, and interact with the world.

The invention of the printing press in the 15th century democratized knowledge and fueled the Renaissance. This groundbreaking technology enabled the mass production of books, making information accessible to a broader audience and breaking the monopoly of knowledge held by the elite. The printing press facilitated the spread of new ideas, scientific discoveries, and literary works, fostering a culture of intellectual curiosity and critical thinking that transformed European society.

The telescope's development in the 17th century revolutionized our understanding of the universe. Galileo Galilei's observations of the moon, planets, and stars challenged the geocentric view of the cosmos and paved the way for the scientific revolution. The telescope extended human vision beyond our planet's confines, revealing the universe's vastness and inspiring new questions about our place in the cosmos. This newfound knowledge fueled scientific inquiry and led to groundbreaking discoveries in astronomy and physics.

The invention of the steam engine in the 18th century powered the Industrial Revolution, transforming manufacturing, transportation, and agriculture. The steam engine's ability to convert heat energy into mechanical work led to the development of factories, railways, and steam-powered ships, dramatically increasing productivity and facilitating the movement of goods and people. This technological advancement spurred economic growth, urbanization, and the rise of industrial societies.

The invention of the telegraph in the 19th century revolutionized communication, enabling near-instantaneous transmission of messages across vast distances. The telegraph connected communities, facilitated trade, and accelerated the dissemination of news and information. This breakthrough in communication technology fostered a sense of global interconnectedness and laid the groundwork for developing the telephone, radio, and the internet.

These are just a few examples of how mechanical marvels have transformed human intelligence and shaped history. These innovations have propelled us toward a smarter, more interconnected, and technologically advanced society by extending our capabilities, expanding our knowledge, and connecting us in unprecedented ways.

WHISPERS IN THE SHADOWS: THE ANCIENT ROOTS OF ENCRYPTION

Encryption has emerged as a cornerstone of modern cybersecurity in the relentless pursuit of safeguarding secrets and ensuring the secure exchange of information. Nevertheless, the roots of this practice stretch far back into the annals of human history, entwined with the very emergence of written communication. This chapter embarks on a journey through time, exploring the intriguing history of encryption in early human societies, where the desire to protect sensitive information from prying eyes gave rise to ingenious techniques for concealing messages and ensuring their confidentiality.

The need to protect those inscriptions from unauthorized access arose from the moment humans began inscribing their thoughts and knowledge. Ancient civilizations, with their complex social structures, political intrigues, and military strategies, recognized the importance of keeping sensitive information hidden from prying eyes. Whether it was the coded messages of pharaohs in ancient Egypt, the cryptic missives of Roman generals, or the secret diplomatic communiqués of Mesopotamian kings, the desire to safeguard secrets fueled the development of ingenious techniques for concealing messages and ensuring their confidentiality.

This chapter delves into the fascinating world of ancient encryption, exploring the creative methods employed by our ancestors to protect their secrets. We will uncover the hidden messages embedded in hieroglyphs and cuneiform script, decipher the ingenious ciphers used by Spartan warriors, and unravel the secrets concealed within the intricate patterns of ancient textiles.

Through this exploration, we will discover that the pursuit of secrecy is not a modern invention but rather an enduring human endeavor woven into the very fabric of our history. The techniques developed by early civilizations, though seemingly simple compared to the complex algorithms of today, laid the foundation for the sophisticated cybersecurity systems that protect our digital world. By tracing the evolution of encryption from its ancient roots to its modern manifestations, we can gain a deeper appreciation for the ingenuity and resourcefulness of those who sought to safeguard information and ensure the secure exchange of knowledge across the ages.

THE DAWN OF SECRECY: EARLY FORMS OF ENCRYPTION

The earliest forms of encryption emerged from the cradle of civilization, entwined with the development of writing systems and the burgeoning need to safeguard sensitive information. In ancient societies, where knowledge was power and secrets held immense strategic value, rulers, military leaders, and merchants sought ingenious ways to protect their communications from prying eyes. They understood that the confidentiality of their messages, whether military orders, diplomatic dispatches, or trade secrets, could mean the difference between victory and defeat, prosperity and ruin.

One of the earliest known examples of encryption, the Caesar cipher, is attributed to Julius Caesar himself, a testament to the enduring importance of information security in power. This deceptively simple substitution cipher involved shifting each alphabet letter by a fixed number of positions, transforming a readable message into an indecipherable jumble of characters. Only those possessing the secret key – knowing how many positions to shift the letters back – could unlock the true meaning hidden within the apparent gibberish.

Another fascinating example of ancient encryption ingenuity is the scytale cipher employed by the Spartans, a society renowned for its military prowess and discipline. This method involved wrapping a parchment strip around a specific diameter rod. The message was then written across the parchment, each letter falling on a different strip turn. When unwrapped, the message appeared as a seemingly random sequence of letters, a puzzle that could only be solved by those possessing a rod of the same diameter, allowing them to reconstruct the original message.

These early encryption techniques, though simple by today's standards, represent a profound understanding of the importance of confidentiality and the ingenuity of our ancestors in devising methods to protect sensitive information. They laid the foundation for the complex and sophisticated encryption algorithms safeguard our digital communications today. This is a testament to the enduring human desire to shield knowledge from unauthorized access and ensure the secure exchange of information in an ever-evolving world.

ENCRYPTION IN ANCIENT EGYPT AND MESOPOTAMIA

In the cradle of civilization, along the fertile banks of the Nile and between the Tigris and Euphrates rivers, ancient Egypt and Mesopotamia civilizations cultivated not only agriculture and monumental architecture but also the art of secrecy. Their scribes, the guardians of knowledge and written communication, developed ingenious techniques to protect sensitive information from prying eyes. Egyptian scribes, masters of hieroglyphic writing, employed intricate substitutions and transpositions, transforming their elegant pictorial script into enigmatic puzzles that concealed messages of political and religious significance. Meanwhile, in the bustling city-states of Mesopotamia, cuneiform-based ciphers were developed to safeguard diplomatic dispatches and military strategies, ensuring that the intentions of rulers and generals remained hidden from their adversaries.

However, the ingenuity of these ancient civilizations extended beyond simple ciphers and codes. They also mastered the art of steganography, the subtle art of hiding messages within seemingly innocuous communications, like whispers concealed within a bustling marketplace. Messages were woven into the very fabric of their culture, concealed within intricate patterns on textiles, etched onto the underside of clay tablets, or rendered invisible to the naked eye with special inks. These techniques ensured that the very existence of a secret message remained shrouded in mystery, safeguarding sensitive information from those who lacked the knowledge or tools to unveil its hidden presence.

This mastery of secrecy, born from the desire to protect power, knowledge, and strategic advantage, laid the foundation for the sophisticated encryption techniques that underpin our modern digital world. The ingenuity of these ancient civilizations serves as a testament to the enduring human drive to safeguard information and ensure secure communication, a drive that continues to shape our interconnected world today.

THE EVOLUTION OF ENCRYPTION TECHNIQUES

As societies grew and the complexity of communication evolved, so did the sophistication of encryption techniques. The need to safeguard sensitive information, whether military secrets, diplomatic dispatches, or financial transactions, spurred the development of increasingly complex ciphers. These intricate systems, involving multiple layers of substitutions, transpositions, and even mathematical operations, presented formidable challenges to those seeking to decipher intercepted messages.

The invention of the Vigenère cipher in the 16th century marked a significant leap forward in encryption technology. This polyalphabetic cipher, a masterpiece of cryptographic ingenuity, employed a keyword to encrypt messages, adding a layer of complexity that confounded traditional cryptanalysis techniques. Unlike simpler substitution ciphers, where another letter consistently replaces each letter, the Vigenère cipher used a shifting key, effectively encrypting each letter with a different substitution alphabet. This made it resistant to frequency analysis, a standard method codebreakers use to exploit the predictable patterns of letter frequencies in natural language.

The Vigenère cipher, with its intricate key-based encryption, remained a formidable challenge for centuries, effectively safeguarding sensitive communications and playing a crucial role in military and diplomatic affairs. Its invention marked a turning point in the history of cryptography, demonstrating the power of human ingenuity to protect information and ensure secure communication in an increasingly complex world.

THE LEGACY OF ANCIENT ENCRYPTION

The ingenuity and resourcefulness of early humans in developing encryption techniques laid the foundation for the sophisticated cybersecurity systems underpinning our modern digital world. These ancient innovators, driven by the need to protect sensitive information and ensure secure communication, devised clever methods to conceal messages and safeguard secrets from prying eyes. The fundamental principles they pioneered, such as substitution, where letters or symbols are replaced with others according to a secret key, and transposition, where the order of characters is rearranged, remain cornerstones of encryption in the digital age.

While modern encryption algorithms' complexity and computational power dwarf those of their ancient counterparts, the underlying principles remain remarkably similar. The Caesar cipher, employed by Julius Caesar to protect military communications, involved a simple shift of letters in the alphabet, a basic form of substitution that foreshadows the complex algorithms used to secure our online transactions today. The scytale cipher, used by the Spartans, relied on the physical transposition of letters on a strip of parchment wrapped around a rod, a precursor to the intricate mathematical transformations that underpin modern encryption techniques.

The enduring legacy of these early encryption efforts is a powerful reminder that the desire to protect information and ensure secure communication is not a modern invention but an intrinsic human drive that has persisted throughout history. From the simple ciphers used by ancient rulers to safeguard military secrets to the complex algorithms that protect our digital communications today, the pursuit of secrecy has been an unbroken thread woven into the very fabric of human civilization. This pursuit reflects a desire for privacy and security and a fundamental understanding of the power of information and the importance of safeguarding it from those who would seek to exploit it.

THE DIGITAL DAWN: ELECTRONICS, CONNECTIVITY, AND THE RESHAPING OF HUMAN INTELLIGENCE

The human story is a fabric with threads of innovation, adaptation, and an unyielding quest for progress. Throughout history, moments of profound transformation have punctuated this narrative, propelling us into new eras of understanding and possibility. The invention of electronics and the subsequent rise of the digital world marks one such turning point, a revolution that has reshaped the landscape of human intelligence and societal interaction. This chapter embarks on a journey to explore the profound impact of these technological advancements, tracing their origins from the spark of discovery to their pervasive influence on our cognitive development and their pivotal role in shaping the interconnected world we inhabit today.

Like a bolt of lightning illuminating the night sky, the invention of electronics ignited a technological revolution that would forever alter the course of human civilization. From the humble beginnings of the vacuum tube and the transistor, these innovations harnessed the power of electrons to control and amplify electrical signals, laying the foundation for a cascade of inventions that would transform communication, computation, and information processing. The telegraph and telephone, early triumphs of electronic ingenuity, transcended geographical boundaries, connecting individuals and communities in ways never imagined. The radio, with its ability to transmit sound wirelessly, brought news, entertainment, and cultural experiences into homes across the globe, fostering a sense of shared experience and global interconnectedness.

The invention of the microchip in the mid-20th century marked a paradigm shift in electronics, miniaturizing complex electronic circuits and enabling the mass production of powerful computing devices. This breakthrough paved the way for the development of personal computers, smartphones, and the internet, ushering in the digital age and fundamentally transforming the way we live, work, and interact with each other. The digital revolution, fueled by the exponential growth of computing power and the interconnectedness of the internet, has democratized access to information, empowered individuals, and created a global village where knowledge flows freely and instantaneously across borders and cultures.

This unprecedented connectivity has fostered collaboration, innovation, and cultural exchange on an unprecedented scale, accelerating the pace of technological advancement and reshaping human intelligence in profound ways. Accessing vast amounts of information at our fingertips has expanded our knowledge base, fostered new forms of learning, and empowered us to solve complex problems with unprecedented speed and efficiency. However, this digital deluge also presents new challenges for human cognition. The constant bombardment of information and stimuli can lead to information overload, distraction, and a decline in critical thinking skills. Navigating the digital landscape requires new cognitive tools and strategies to effectively filter, analyze, and synthesize information.

The transformative power of electronics and digital technology is evident in countless examples. The personal computer, once a bulky and expensive machine confined to research labs and universities, has become an indispensable tool for creativity, communication, and productivity, empowering individuals and fueling innovation across all sectors of society. The internet, a vast network of interconnected computers, has broken down geographical barriers, facilitating communication, collaboration, and the sharing of knowledge on a global scale. The smartphone, a ubiquitous symbol of the digital age, has put powerful computing devices in the hands of billions, transforming how we access information, communicate, and navigate our daily lives.

As we navigate the ever-evolving digital landscape, harnessing these technologies' power for humanity's betterment is crucial. By fostering critical thinking, promoting digital literacy, and ensuring equitable access to technology, we can empower individuals and communities to thrive in the digital age. The invention of electronics and the rise of the digital world have ushered in a new era of human intelligence and societal advancement, with challenges and opportunities. By embracing the transformative potential of these technologies while remaining mindful of their impact on our cognitive development and social structures, we can shape a future where the digital revolution catalyzes a more informed, interconnected, and equitable world.

THE SPARK OF INNOVATION: THE BIRTH OF ELECTRONICS

The late 19th and early 20th centuries witnessed the germination of the digital revolution, a period marked by groundbreaking innovations that would forever alter the trajectory of human communication and technological progress. At the heart of this transformation lay the invention of the vacuum tube, a device that harnessed the power of electrons to control and amplify electrical signals. This pioneering technology, though bulky and inefficient by today's standards, paved the way for the development of radios, televisions, and early computers, revolutionizing how information was transmitted and processed.

Building upon the foundation laid by the vacuum tube, the invention of the transistor in 1947 marked a paradigm shift in electronics. This revolutionary device, smaller, more efficient, and vastly more reliable than its predecessor, ushered in the era of miniaturization and mass production of electronic circuits. The transistor's impact was nothing short of transformative, enabling the development of portable radios, compact televisions, and, most importantly, the microprocessors that would power the personal computer revolution.

These early innovations, though seemingly simple in retrospect, unleashed a cascade of technological advancements that continue to reshape our world today. The ability to control and manipulate electrical signals with increasing precision and efficiency laid the groundwork for developing sophisticated electronic devices that have transformed communication, computation, and information processing. From the smartphones in our pockets to the supercomputers that drive scientific discovery, the legacy of these early electronic pioneers continues to shape the digital landscape and propel us toward a future of unprecedented technological possibility.

THE DIGITAL REVOLUTION: FROM VACUUM TUBES TO MICROCHIPS

The invention of the microchip in the mid-20th century was nothing short of a technological earthquake, its tremors forever altering the landscape of human existence. This tiny marvel of engineering, a sliver of silicon etched with intricate circuitry, unleashed a paradigm shift in electronics, enabling the miniaturization and mass production of complex electronic systems that were previously unimaginable. With its ability to pack immense computational power into ever-shrinking dimensions, the microchip became the beating heart of a digital revolution that would transform how we live, work, and interact with each other.

This breakthrough ignited an explosion of innovation, paving the way for the development of personal computers that brought computing power into homes and offices, liberating it from the confines of bulky mainframes and research labs. The microchip's versatility extended beyond computation, enabling the creation of smartphones that have become indispensable extensions of ourselves, connecting us to a global network of information and communication. The internet, a vast interconnected web of digital information, emerged as a direct consequence of the microchip's power, transforming how we access knowledge, share ideas, and build communities.

The digital age, ushered in by the microchip, has fundamentally reshaped the human experience. We now inhabit a world where information flows freely and instantaneously across borders and cultures, communication transcends geographical limitations, and knowledge is readily available at our fingertips. This interconnectedness has fostered collaboration, accelerated innovation, and democratized access to information, empowering individuals and communities in unprecedented ways.

The microchip's impact extends far beyond technological advancement; it has fundamentally altered the fabric of human society. It has transformed the workplace, automating tasks, increasing productivity, and creating new forms of employment. It has revolutionized education, providing access to vast knowledge libraries and enabling new learning modes. It has reshaped social interactions, connecting people across continents and fostering new community and cultural exchange forms.

In essence, the cyber invention of the microchip has been a catalyst for a profound transformation of human intelligence and societal organization. It has expanded our cognitive horizons, accelerated the pace of innovation, and created a world where connectivity and information sharing have become defining features of human experience. As we continue to navigate the ever-evolving digital landscape, the microchip's legacy serves as a reminder of the transformative power of human ingenuity and the boundless possibilities that emerge when we dare to push the boundaries of technological innovation.

CONNECTIVITY AND THE RISE OF THE GLOBAL VILLAGE

The digital revolution, ignited by the invention of the microchip and propelled by the pervasive interconnectedness of the internet, has fundamentally reshaped the human experience, shrinking our world into a global village where information traverses borders and cultures with unprecedented speed and freedom. This interconnectedness, a defining characteristic of the digital age, has fostered collaboration, knowledge sharing, and cultural exchange on a scale never imagined. The barriers of distance and time have crumbled, allowing individuals from all corners of the globe to connect, communicate, and collaborate in real time. This seamless exchange of ideas and information has accelerated the pace of innovation across all fields of human endeavor, from science and technology to art and culture. Once fragmented and isolated, humanity's collective intelligence is now increasingly interwoven, forming a dynamic and evolving network of minds that drives progress and reshapes our understanding of the world.

THE IMPACT ON COGNITIVE DEVELOPMENT

With its ceaseless torrent of information and stimuli, the digital world has undeniably revolutionized the landscape of human intelligence, presenting unprecedented opportunities and formidable challenges. Accessing vast knowledge repositories at our fingertips has democratized information, empowering individuals with a breadth and depth of previously unimaginable understanding. This readily available information has fostered new forms of learning, enabling self-directed exploration and collaborative problem-solving on a global scale. However, this digital deluge also presents a formidable challenge to our cognitive faculties. The constant bombardment of notifications, updates, and fleeting trends can fragment our attention, hindering our ability to focus and engage in deep, sustained thought. The ephemeral nature of digital information, often lacking the context and vetting of traditional sources, can also undermine critical thinking skills, making it challenging to discern fact from fiction in the swirling sea of online content. Moreover, the sheer volume of information can lead to a sense of overwhelm, a cognitive overload that hinders our ability to process, synthesize, and make meaningful connections.

REAL-WORLD EXAMPLES OF TRANSFORMATION

The transformative power of electronics and digital technology is inextricably woven into the fabric of modern life, evident in the myriad innovations that have dramatically reshaped human intelligence and the very structures of our societies. Consider the invention of the personal computer, a revolutionary leap in accessible computing power that democratized technology and placed the power of computation into the hands of individuals. This democratization ignited a surge of creative expression, scientific exploration, and entrepreneurial endeavors, empowering individuals to transform their ideas into tangible realities. Suddenly, the power to create groundbreaking software applications, design innovative tools, produce captivating artistic creations, and develop transformative educational resources was within reach, not just for large institutions or corporations but for individuals driven by their ingenuity and vision. This shift unleashed a wave of innovation that continues to ripple through every facet of modern life.

The rise of the internet, a vast interconnected network spanning the globe, has dissolved geographical barriers and fostered a global community, bringing people together in ways never imagined. Once limited by distance and time, communication has been revolutionized, allowing for instantaneous exchange of ideas and information across continents. This unprecedented connectivity has transformed collaboration, enabling researchers, artists, and activists to collaborate seamlessly, regardless of physical location.

The internet has profoundly impacted the sharing of knowledge. Educational resources, research findings, and cultural expressions are now readily available to anyone with an internet connection, democratizing access to information and empowering individuals with the tools for lifelong learning. This interconnectedness has fueled scientific breakthroughs, as researchers worldwide can collaborate on complex problems, share data, and accelerate the pace of discovery.

Artistic collaborations have flourished in the digital age, with musicians, writers, and visual artists joining forces across borders to create new and innovative forms of expression. Social movements have also harnessed the power of the internet to mobilize support, organize protests, and advocate for change on a global scale. The internet has become a powerful platform for amplifying marginalized voices, challenging oppressive systems, and fostering a sense of shared purpose and collective action.

In essence, the rise of the internet has woven a digital tapestry that connects individuals, communities, and cultures in a way that transcends geographical boundaries. This interconnectedness has revolutionized communication and collaboration and empowered individuals and communities to learn, create, and advocate for change unimaginably.

The development of smartphones, those ubiquitous pocket-sized powerhouses, has fundamentally reshaped the human experience, blurring the lines between the physical and digital realms to an unprecedented degree. These devices have transcended their initial role as mere communication tools, evolving into extensions of ourselves, seamlessly integrated into the fabric of our daily lives. They provide instant access to a vast ocean of information, connecting us to the global knowledge network and enabling us to learn about the world with a few taps on a screen. They serve as indispensable communication hubs, facilitating instant connections with loved ones, colleagues, and communities across continents. Furthermore, as navigational aids, they guide us through unfamiliar landscapes, transforming how we explore and experience the world around us. In essence, smartphones have become indispensable tools for navigating the complexities of modern life, empowering us with knowledge, connection, and a newfound sense of agency in a rapidly changing world.

The invention of electronics and the rise of the digital world have undeniably ushered in a new era of human intelligence and societal advancement. The unprecedented connectivity, access to information, and computational power afforded by these technologies have fundamentally altered how we think, learn, and interact with the world. We are constantly interconnected, where information flows freely and instantaneously across geographical boundaries, transforming how we communicate, collaborate, and consume knowledge. This digital revolution has opened up many opportunities for human intelligence to flourish in ways never imagined.

However, this digital landscape is not without its challenges. The constant bombardment of information and stimuli can lead to shorter attention spans, a decline in critical thinking skills, and the potential for information overload. The ease with which misinformation can spread through digital channels poses a significant threat to informed decision-making and societal cohesion.

Despite these challenges, the digital revolution has fostered new creativity, collaboration, and problem-solving forms. Online platforms and virtual communities have become fertile ground for innovation, where individuals from diverse backgrounds can connect, share ideas, and collaborate on projects that transcend geographical limitations. Accessing vast information and computational resources has empowered individuals and communities to tackle complex problems, from scientific research and technological development to social advocacy and artistic expression.

The digital age presents a complex and dynamic landscape for human intelligence that demands adaptability, critical thinking, and a mindful approach to technology consumption. By harnessing the power of digital tools while remaining aware of their potential pitfalls, we can navigate this evolving landscape and ensure that the digital revolution continues to catalyze human progress and societal advancement.

As we navigate the ever-shifting terrain of the digital revolution, it becomes increasingly crucial to harness these technologies' immense power for humanity's betterment. This necessitates a conscious and proactive approach, ensuring technological advancements are tools for progress, inclusivity, and empowerment. By fostering critical thinking, we equip individuals to discern truth from falsehood in the vast ocean of online information, evaluate evidence with a discerning eye, and engage in thoughtful discourse that transcends echo chambers and filter bubbles.

Promoting local and global collaboration allows us to leverage the interconnectedness of the digital world to address shared challenges, from climate change and global health crises to social inequalities and economic disparities. By breaking down barriers and fostering open dialogue, we can harness the collective intelligence of diverse communities to find innovative solutions and build a more sustainable and equitable future.

Ensuring equitable access to technology is paramount, bridging the digital divide that separates those with access to information and opportunities from those without. By providing affordable internet access, digital literacy training, and the necessary infrastructure to underserved communities, we can empower individuals, foster economic growth, and create a more inclusive digital society.

In essence, the digital revolution presents a unique opportunity to shape a world that is more informed and interconnected, and more just and equitable. By embracing the values of critical thinking, collaboration, and inclusivity, we can ensure that technology catalyzes positive change, empowering individuals, strengthening communities, and fostering a future where all share the benefits of the digital age.

2 The Hyper-Connected World

Human Intelligence in the Age of Blurred Realities

The 21st century has ushered in an era of unprecedented interconnectedness, where the boundaries between physical and digital realities are becoming increasingly blurred. This chapter explores the profound impact of this hyper-connected world on human intelligence, examining the challenges and opportunities presented by the rapid advancements in technology, the rise of artificial intelligence (AI), and the growing reliance on virtual interactions.

The convergence of technologies like high-speed internet, ubiquitous mobile devices, and pervasive social media platforms has woven a dense web of interconnectedness, shrinking geographical distances and fostering instant global communication. This hyper-connectivity has transformed how we interact, learn, and conduct business, creating a global village where information flows freely and individuals can connect with others regardless of physical location.

This hyper-connected world presents both unprecedented opportunities and complex challenges for human intelligence. On the one hand, the ease of access to information, the ability to connect with diverse perspectives, and the collaborative potential of online platforms have the potential to enhance cognitive skills, foster creativity, and accelerate innovation. On the other hand, the constant bombardment of information, the blurring of real and virtual interactions, and the growing reliance on artificial intelligence raise concerns about attention spans, critical thinking skills, and the potential for information overload.

The rise of AI is a defining feature of the hyper-connected world. AI-powered systems are increasingly integrated into our lives, from personalized recommendations and automated tasks to medical diagnoses and financial trading. While AI can enhance human capabilities and solve complex problems, it raises ethical concerns about job displacement, algorithmic bias, and the potential for misuse.

The COVID-19 pandemic accelerated the shift toward virtual interactions, with remote work, online education, and digital social gatherings becoming increasingly common. While virtual interactions offer convenience and flexibility, they also raise questions about the impact on social skills, emotional well-being, and the development of meaningful relationships.

As we navigate this hyper-connected world, we must develop a critical understanding of the technologies that shape our lives and the potential impact on human

DOI: 10.1201/9781003641506-2

intelligence. This includes fostering digital literacy, promoting responsible technology use, and addressing ethical concerns about AI and data privacy.

The hyper-connected world presents a complex and dynamic landscape for human intelligence. The unprecedented connectivity, access to information, and blurring of real and virtual realities offer immense opportunities for collaboration, innovation, and societal advancement. However, the challenges of information overload, the ethical implications of AI, and the potential impact on social and cognitive skills demand careful consideration and proactive measures to ensure that the hyper-connected world catalyzes a more informed, equitable, and human-centered future.

THE RISE OF THE SUPER-CONNECTED WORLD

The 21st century has ushered in an era of unprecedented interconnectedness, where the once vast expanse of our planet has shrunk into a global village, pulsating with the rhythm of instant communication and rapid transit. Supersonic air travel has collapsed geographical distances, enabling the traverse of continents in hours. At the same time, high-speed internet and satellite communication have woven an intricate web of information exchange, connecting individuals and communities across the globe in ways never before imagined. This convergence of technologies has created a tapestry of human experience that transcends geographical boundaries and cultural divides, fostering a sense of shared existence and global citizenship.

This chapter delves into the transformative power of this hyperconnectivity, exploring its profound impact on human interaction, cultural exchange, and the pursuit of a truly interconnected global society. We examine how the ability to communicate instantaneously across vast distances has reshaped social relationships, business practices, and political discourse. We explore how the internet and social media have facilitated the exchange of cultural expressions, ideas, and perspectives, fostering a greater understanding and appreciation of human diversity.

Furthermore, we investigate the role of hyperconnectivity in driving collaborative efforts to address global challenges, from climate change and pandemics to poverty and inequality. We examine the emergence of international organizations and initiatives dedicated to bridging the digital divide, promoting digital literacy, and ensuring equitable access to technology for all.

This chapter also delves into the potential downsides of hyperconnectivity, exploring the challenges of misinformation, online privacy, and the erosion of traditional cultural values in the face of globalization. By examining both the opportunities and challenges presented by this interconnected world, we aim to provide a comprehensive understanding of its transformative power and its implications for the future of human society.

THE DAWN OF THE GLOBAL VILLAGE

The dream of a global village, where individuals and communities from all corners of the world could seamlessly connect and interact, once a mere glimmer in the human imagination, is rapidly materializing. Supersonic air travel has shrunk the vast expanse of our planet, compressing the time it takes to traverse continents from days to hours. This accelerated mobility has unlocked unprecedented opportunities for face-to-face interactions, fostering cultural exchange, collaborative endeavors, and forging personal bonds that transcend geographical boundaries.

High-speed internet, a ubiquitous and indispensable tool of the modern era, has dissolved distance barriers, enabling instantaneous communication and collaboration across vast geographical expanses. With the click of a button or the touch of a screen, we can connect with colleagues, friends, and family members on the other side, sharing ideas, experiences, and emotions in real-time. This seamless flow of information has fostered a sense of global interconnectedness, facilitating knowledge exchange, coordinating efforts, and forming virtual communities that transcend national borders.

Satellite communication, extending its reach to even the most remote corners of the globe, has played a crucial role in bridging the digital divide, bringing connectivity to underserved communities and fostering a sense of inclusivity in the global village. This expanded access to information and communication technologies (ICTs) has empowered individuals, facilitated education, and spurred economic development in regions isolated from the digital mainstream.

The convergence of these technological advancements has woven a tapestry of interconnectedness, creating a world where physical distance no longer dictates the limits of human interaction and collaboration. With its vibrant mix of cultures, perspectives, and ideas, this global village holds immense potential for fostering understanding, promoting cooperation, and driving innovation on a global scale.

The ease of travel and communication has woven a vibrant tapestry of human interaction and collaboration, transcending geographical boundaries and cultural differences. Businesses now operate seamlessly across continents, their teams spanning diverse time zones and linguistic backgrounds, united by a shared digital workspace. Researchers collaborate on global challenges, sharing data and insights in real-time, accelerating the pace of discovery and innovation. Individuals connect with loved ones across vast distances, maintaining intimate relationships and sharing experiences with ease. This interconnectedness has fueled economic interdependence as global markets intertwine and supply chains stretch across continents. Cultural exchange flourishes as ideas, art, and traditions are shared and celebrated across borders, enriching our understanding of the human experience in all its diversity. The rapid dissemination of information empowers individuals with knowledge, fosters critical thinking, and fuels democratic participation on a global scale. In this interconnected world, knowledge, ideas, and innovation flow freely, transcending the limitations of physical distance and fostering a shared sense of humanity.

THE ROLE OF INTERNATIONAL ORGANIZATIONS

Recognizing the transformative power of connectivity, international organizations are shaping a future where digital inclusion and equitable access to information become cornerstones of a more just and interconnected world. Through its ambitious Sustainable Development Goals, the United Nations has spotlighted the importance of universal access to ICTs. These technologies are no longer viewed as mere conveniences but as powerful enablers of sustainable development, with the potential to transform education, healthcare, economic opportunity, and social progress across the globe.

Beyond simply providing access, these organizations are grappling with the complex challenges of internet governance, striving to create a digital landscape that is both inclusive and secure. The International Telecommunication Union (ITU), a specialized agency of the United Nations focused on ICTs, is working to harmonize global standards, promote innovation, and bridge the digital divide that separates those with access to technology from those without.

The World Bank, recognizing the crucial role of connectivity in fostering economic growth and reducing poverty, is investing in infrastructure development, digital literacy programs, and policy reforms to ensure that the benefits of the digital revolution are shared equitably across all nations. These efforts are not merely about expanding access to the internet but about fostering a truly inclusive digital ecosystem where everyone, regardless of their background or location, can participate in and benefit from the opportunities of the digital age.

Recognizing the transformative power of connectivity, international organizations are shaping a future where digital inclusion and equitable access to information become cornerstones of a more just and interconnected world. Through its ambitious Sustainable Development Goals, the United Nations has spotlighted the importance of universal access to ICTs. These technologies are no longer viewed as mere conveniences but as powerful enablers of sustainable development, with the potential to transform education, healthcare, economic opportunity, and social progress across the globe. For example, the UN's Technology Bank for Least Developed Countries provides technical assistance and capacity building to help these nations bridge the digital divide and leverage technology for sustainable development.

Beyond simply providing access, these organizations are grappling with the complex challenges of internet governance, striving to create a digital landscape that is both inclusive and secure. The International Telecommunication Union, a specialized agency of the United Nations focused on information and communication technologies, is working to harmonize global standards, promote innovation, and bridge the digital divide that separates those with access to technology from those without. This includes initiatives such as the Connect 2030 Agenda, which sets targets for universal and affordable access to ICTs, and the Global Cybersecurity Agenda, which promotes international cooperation to address cyber threats.

The World Bank, recognizing the crucial role of connectivity in fostering economic growth and reducing poverty, is investing in infrastructure development, digital literacy programs, and policy reforms to ensure that the benefits of the digital revolution are shared equitably across all nations. This includes projects like the Digital Economy for Africa initiative, which aims to expand digital access and accelerate digital transformation across the continent.

However, achieving a truly connected world is not without its challenges. Censorship, digital surveillance, and the uneven distribution of technological resources pose significant obstacles. Overcoming these challenges will require continued collaboration and investment in digital inclusion initiatives, ensuring that the digital age empowers all individuals and communities, regardless of their background or location.

CHALLENGES TO ACHIEVING A TRULY INTERCONNECTED SOCIETY

While the rise of hyperconnectivity offers immense promise for a more interconnected and collaborative world, significant challenges remain in achieving a truly global society where information and opportunities are shared equitably. One of the most pressing challenges is the uneven distribution of technology, with stark disparities in access to the internet and digital devices across different regions and socioeconomic groups. This digital divide creates a chasm, excluding marginalized communities from the benefits of the digital age and perpetuating existing inequalities.

Furthermore, the censorship practices employed by some countries pose a significant barrier to the free flow of information and ideas, hindering the development of a truly connected global society. By restricting access to certain websites, social media platforms, and online content, these governments create a digital iron curtain, limiting freedom of expression, stifling innovation, and preventing the cross-pollination of ideas that fuel progress and understanding. This harms the citizens within those countries and fragments the global information landscape, hindering the collective pursuit of knowledge and collaborative problem-solving.

A multifaceted and globally collaborative approach is essential to fully harness the transformative potential of hyperconnectivity and navigate its inherent complexities. Bridging the digital divide, that chasm that separates those with access to technology and its accompanying opportunities from those without, necessitates a concerted investment in infrastructure development, digital literacy education, and policies that promote equitable access for all. This includes expanding broadband internet access to underserved communities, providing affordable devices and training programs, and fostering an inclusive digital environment where everyone can participate and contribute.

Combating censorship, a pervasive threat to the free flow of information and the exercise of fundamental human rights, requires a united front on the international stage. Governments, organizations, and individuals must advocate for digital rights, challenge repressive policies, and support developing and deploying technologies that circumvent censorship and empower individuals to access and share information freely. This includes promoting encryption technologies, anonymization tools, and decentralized platforms that resist centralized control and foster open communication.

Furthermore, addressing the ethical considerations and potential risks associated with hyperconnectivity is crucial. This involves fostering critical thinking skills, promoting media literacy, and engaging in open dialogue about the impact of technology on our lives, societies, and values. By proactively addressing these challenges, we can ensure that hyperconnectivity catalyzes positive change, fostering a more informed, interconnected, and equitable world.

The hyper-connected world, woven by the intricate threads of the internet, mobile technology, and instant communication, presents a landscape brimming with unprecedented opportunities and formidable challenges. This interconnectedness has the potential to revolutionize human interaction, democratize knowledge, and foster collaboration on a global scale. Imagine a world where geographical barriers dissolve, where individuals from all walks of life can readily access information and contribute to the global exchange of knowledge and ideas. This vision, however, is not without its obstacles.

The digital divide, the chasm that separates those with access to technology and those without, threatens to exacerbate existing inequalities and create new forms of exclusion. Censorship, the suppression of information and freedom of expression, casts a shadow over the digital landscape, hindering the free flow of knowledge and stifling innovation. The rise of misinformation and the manipulation of online narratives threaten informed decision-making and the foundations of trust in the digital age.

We can foster a more interconnected, informed, and collaborative global society by embracing the transformative power of connectivity while simultaneously addressing the barriers that hinder its equitable distribution. This requires a multipronged approach, encompassing investments in infrastructure to bridge the digital divide, developing robust cybersecurity measures to protect against malicious actors, and promoting digital literacy to empower individuals to navigate the complexities of the online world critically and responsibly.

Pursuing a truly connected world, where information flows freely, and individuals from all walks of life can participate in the global exchange of knowledge and ideas, remains a defining challenge and opportunity of our time. It is a pursuit that demands technological innovation and a steadfast commitment to the values of openness, inclusivity, and the free exchange of ideas. By embracing these values and actively working to overcome the barriers that hinder connectivity, we can harness the transformative power of the digital age to build a more informed, equitable, and interconnected world for all.

CHALLENGES TO HUMAN INTELLIGENCE

The rise of the hyper-connected world, a tapestry woven with the threads of instant communication, global collaboration, and boundless information access, has ushered in an era of unprecedented opportunity and unparalleled challenges for human intelligence. This chapter plunges into the intricate dance between connectivity, cognition, and societal development, exploring the transformative potential of a world where borders blur and knowledge flows freely while acknowledging the persistent barriers that hinder the full realization of this interconnected dream.

We stand at a crossroads where the digital symphony of interconnected networks amplifies humanity's collective intelligence. The ability to share ideas, collaborate on projects, and access information from around the globe has ignited a spark of innovation, accelerating scientific breakthroughs, artistic collaborations, and social movements on an unprecedented scale. Nevertheless, this hyper-connectivity also presents a unique set of challenges, testing the very fabric of human cognition and societal structures.

The constant bombardment of information, the blurring of real and virtual interactions, and the growing reliance on artificial intelligence raise profound questions about the future of human attention, critical thinking, and even our sense of identity. As we navigate this complex landscape, we must grapple with the ethical implications of emerging technologies, the potential for digital divides to exacerbate existing inequalities, and the ever-present threat of cyberattacks that can disrupt critical infrastructure and undermine the trust that binds our interconnected world.

This chapter embarks on a journey through the multifaceted dimensions of this hyper-connected age, exploring the transformative potential of a world without borders while acknowledging the barriers that persist. We will delve into the cognitive implications of constant connectivity, examining how our brains adapt to the digital deluge and the potential impact on attention, memory, and critical thinking. We will explore the societal implications of a world where physical boundaries dissolve, examining the opportunities for global collaboration and cultural exchange while addressing the challenges of digital divides, censorship, and the erosion of privacy.

Ultimately, this chapter seeks to illuminate the path toward a future where the hyper-connected world catalyzes human flourishing, fostering a global community that is more informed and interconnected, equitable, and resilient.

The constant flow of information, the blurring of real and virtual interactions, and the increasing reliance on artificial intelligence have created a dynamic and ever-shifting cognitive landscape, a swirling vortex of stimuli and interactions that challenge the foundation of human intelligence. While offering unprecedented opportunities for knowledge sharing, collaboration, and cultural exchange, this hyper-connectivity also presents formidable challenges to our cognitive processes, social structures, and even our sense of self.

The deluge of information bombarding us from countless digital sources can overwhelm our cognitive filters, making it difficult to distinguish between reliable sources and misinformation, between meaningful connections and fleeting distractions. The boundaries between real and virtual interactions are becoming increasingly porous, with social media platforms, virtual worlds, and augmented reality (AR) experiences blurring the lines between physical presence and digital representation. This blurring can lead to disorientation, a fragmentation of identity, and a questioning of the authenticity of human connection in a world where technology increasingly mediates reality.

Moreover, the growing reliance on artificial intelligence raises fundamental questions about the future of human cognition. As AI systems become more sophisticated and capable of performing tasks that were once the exclusive domain of human intelligence, we must grapple with the implications for our cognitive development. Will we become overly reliant on AI, outsourcing our critical thinking and decision-making abilities to algorithms and machines? Or will we find ways to harness AI's power to augment our intelligence, creating a synergistic partnership that enhances human capabilities and expands the horizons of knowledge?

The hyper-connected world presents a complex and multifaceted challenge to human intelligence, demanding adaptability, critical thinking, and a conscious effort to maintain our sense of self amid a digital maelstrom. It is a challenge that requires us to re-evaluate our relationship with technology, foster digital literacy, and cultivate a mindful approach to navigating the digital landscape. Only then can we harness the immense potential of hyper-connectivity while safeguarding the essence of human intelligence and ensuring a future where technology empowers rather than diminishes our cognitive and social capabilities.

On one hand, the abundance of information and the ease of communication foster new forms of learning, collaboration, and problem-solving. The ability to access vast knowledge repositories, connect with individuals across geographical boundaries, and engage in real-time discussions with experts in various fields has expanded the horizons of human intelligence, enabling us to tackle complex challenges and generate innovative solutions.

On the other hand, the constant bombardment of information and the fragmented nature of digital communication can lead to cognitive overload, distraction, and a decline in critical thinking skills. Blurring real and virtual interactions can also impact social skills and emotional intelligence as individuals become increasingly accustomed to mediated communication and the curated personas presented online.

THE PARADOX OF CENSORSHIP: FRAGMENTATION IN A CONNECTED WORLD

While technology possesses the remarkable potential to bridge geographical divides and foster a global exchange of ideas, the censorship practices employed by certain countries cast a long shadow over this interconnected world. Often rooted in a desire to control the narrative and suppress dissenting voices, these practices erect barriers to collaboration and hinder the development of a truly connected and intelligent global society.

By restricting access to information and limiting freedom of expression, censorship impedes the free flow of knowledge, stifles innovation, and undermines the very foundations of intellectual progress. It creates an environment where critical thinking is suppressed, diverse perspectives are silenced, and the propagation of dogma replaces the pursuit of truth.

In the realm of technology, censorship limits access to valuable tools and resources and impedes the development of homegrown innovation. When individuals and communities are denied access to the global exchange of ideas, their ability to contribute to the collective pool of knowledge diminishes their technological advancement. It contributes to a widening gap between censored societies and the rest of the world.

Furthermore, censorship contributes to the fragmentation of the global information landscape, creating echo chambers and filter bubbles where individuals are exposed only to information that reinforces their beliefs and biases. This fragmentation hinders cross-cultural understanding, fuels mistrust and animosity, and undermines the potential for collaborative problem-solving on a global scale.

The interconnectedness promised by technology can only be fully realized when information flows freely, and individuals are empowered to express themselves without fear of reprisal. Censorship, in its various forms, undermines this vision, creating a fractured and intellectually impoverished world where the full potential of human ingenuity remains unrealized.

Censorship casts a chilling effect on the free exchange of ideas, stifling the essence of intellectual exploration and open dialogue that fuels societal progress. By depriving individuals of valuable knowledge and diverse perspectives, censorship creates an intellectual vacuum where innovation and creativity struggle to thrive. The advancement of science, technology, and culture hinges on the unfettered flow of information, allowing for the cross-pollination of ideas, the challenging of assumptions, and the collaborative pursuit of solutions to complex problems. When governments restrict access to information, they limit the potential for intellectual growth and impede the progress of society as a whole.

Furthermore, censorship erodes the foundation of trust and openness essential for a connected world. In an era where collaboration and knowledge sharing are crucial for addressing global challenges such as climate change, pandemics, and economic inequality, censorship erects artificial barriers between nations and hinders the collective pursuit of solutions. When governments suppress information and control narratives, they create an atmosphere of suspicion and distrust, undermining the spirit of cooperation essential for navigating an interconnected world's complexities.

The consequences of censorship extend far beyond the immediate silencing of voices and suppression of ideas. It fosters a culture of fear and self-censorship, where individuals hesitate to express dissenting opinions or challenge the prevailing narrative. This chilling effect can stifle creativity, discourage innovation, and ultimately impede the advancement of knowledge and societal progress. In a world where the free flow of information is increasingly recognized as a fundamental human right, censorship is a stark reminder of the fragility of open societies and the constant need to safeguard the principles of freedom of expression and intellectual inquiry.

NAVIGATING THE COMPLEXITIES OF A CONNECTED WORLD

The hyper-connected world of the 21st century presents an exhilarating and daunting landscape for human intelligence. It is a realm of unprecedented interconnectedness, where information flows freely across borders and cultures, and collaboration and innovation flourish at an unprecedented pace. Nevertheless, this interconnectedness also brings forth complex challenges, demanding careful navigation and thoughtful solutions.

The sheer volume of information available in the digital age can be overwhelming, creating a sense of information overload that hinders our ability to discern truth from falsehood, focus on meaningful content, and cultivate deep understanding. The erosion of privacy, with our digital footprints tracked and analyzed by algorithms and corporations, raises concerns about autonomy, self-determination, and the potential for manipulation.

Furthermore, the persistence of censorship, whether imposed by governments or powerful institutions, casts a shadow over the digital landscape, restricting access to information, stifling dissent, and hindering the free exchange of ideas essential for intellectual growth and societal progress.

Navigating this complex terrain requires a multifaceted approach. We must cultivate critical thinking skills to evaluate information with discernment, develop strategies to manage information overload, and advocate for policies that protect privacy and promote digital literacy. We must also challenge censorship in all its forms, championing the free flow of information and fostering a culture of open dialogue and intellectual curiosity.

The hyper-connected world offers immense potential for human advancement, but it also demands a conscious and proactive approach to ensure that technology serves as a tool for empowerment, inclusivity, and the flourishing of human intelligence. By embracing the opportunities and addressing the challenges of this interconnected age, we can shape a future where technology fosters a more informed, just, and equitable world.

As we navigate the ever-evolving digital terrain, a landscape constantly reshaped by technological advancements and societal shifts, it is crucial to cultivate a critical and discerning approach to information and connectivity. Fostering critical thinking skills empowers individuals to sift through the deluge of data, to distinguish fact from fiction, and to engage in thoughtful discourse that transcends the echo chambers and filter bubbles that often confine online interactions.

Promoting digital literacy equips individuals with the tools and knowledge to navigate the digital world safely and effectively, understand the implications of their online actions, and harness technology's power for creative expression, collaboration, and social engagement. Advocating for policies that protect freedom of expression ensures that the digital space remains a vibrant marketplace of ideas where diverse perspectives can be shared, debated, and challenged without fear of censorship or reprisal.

Equitable access to information is paramount, bridging the digital divide that separates those with access to knowledge and opportunity from those without. By ensuring that all share the benefits of the digital revolution, we can foster a more inclusive and just society where everyone can participate in the global conversation and contribute to the collective advancement of human knowledge.

By embracing the transformative potential of connectivity while mitigating its risks, we can harness the power of technology to create a more informed, interconnected, and equitable world. A world where information flows freely, knowledge is shared openly, and the collective intelligence of humanity is unleashed to address the challenges and opportunities of the 21st century and beyond.

Imposed censorship casts a long shadow over society, stifling not only freedom of expression but also the very spirit of intellectual curiosity and open dialogue that fuels progress and innovation. The suppression of ideas, the silencing of dissenting voices, and the distortion of information create a hostile environment where fear and mistrust fester, and a struggle for control replaces the pursuit of knowledge.

In such an environment, cultural resiliency emerges as a powerful force seeking to counter the suffocating effects of censorship. The human desire for freedom of thought and expression finds outlets unexpectedly, often through technologies intended to enforce control. Despite attempts to restrict access and manipulate narratives, the internet has become a fertile ground for exchanging information, organizing resistance, and preserving cultural identity.

The relationship between cultural resiliency and innovation is undeniable. When faced with constraints and limitations, human ingenuity finds ways to circumvent obstacles, adapt, and create new solutions. In its attempt to suppress ideas, censorship inadvertently fuels the creative spirit, driving individuals and communities to seek alternative channels for expression and innovation.

History is replete with examples of this dynamic. The samizdat literature of the Soviet era, clandestinely circulated through underground networks, challenged the official narrative and kept the flame of dissent alive. The rise of independent media and citizen journalism in countries with repressive regimes provides alternative sources of information, countering propaganda and empowering citizens to hold their governments accountable.

Developing encryption technologies and anonymization tools empower individuals to circumvent censorship and protect their digital privacy. The rise of decentralized platforms and blockchain technology offers new avenues for secure communication, information sharing, and collective action organization, challenging centralized authorities' control.

In the face of censorship, cultural resiliency becomes a catalyst for innovation, driving the development of new technologies and strategies to circumvent restrictions and reclaim digital freedoms. This resilience not only preserves cultural identity and fosters critical thinking but also fuels social progress and the pursuit of a more just and equitable society. The struggle against censorship is not merely a battle for freedom of expression; it is a fight for the very soul of a society, its intellectual vitality, and capacity for innovation and progress.

Censorship and a hostile environment can create a breeding ground for adversarial attacks, as the suppression of information and the stifling of dissent foster distrust, paranoia, and a sense of vulnerability. Malicious actors can readily exploit this atmosphere to sow discord, spread misinformation, and manipulate public opinion.

In cybersecurity, censorship can create a false sense of security, leading individuals and organizations to neglect essential safeguards and underestimate the evolving tactics of cyber adversaries. The lack of open dialogue about cyber threats and the suppression of information about vulnerabilities and attack vectors can expose systems and individuals to social engineering and other forms of cyber manipulation.

For instance, in countries with strict internet censorship, the lack of access to information about cybersecurity best practices and emerging threats can make individuals more vulnerable to phishing scams, malware attacks, and online disinformation campaigns. The inability to freely discuss and share information about cyber threats hinders the development of a collective cybersecurity culture, leaving individuals and organizations ill-equipped to defend against increasingly sophisticated attacks.

Furthermore, censorship can create an environment where dissent and critical thinking are suppressed, making it easier for malicious actors to spread propaganda and manipulate public opinion. In such an environment, the lack of diverse perspectives and open debate can lead to a distorted understanding of reality, making individuals more susceptible to manipulation and exploitation.

The chilling effect of censorship can also discourage individuals from reporting cyberattacks or seeking help when they become victims of cybercrime. This fear of reprisal or scrutiny can create a culture of silence, allowing cybercriminals to operate with impunity and further eroding trust in digital systems and institutions.

In conclusion, censorship and a hostile environment can significantly increase the risk of adversarial attacks in both physical and digital realms. The suppression of information, the stifling of dissent, and the erosion of trust create fertile ground for manipulation, exploitation, and misinformation. By fostering open dialogue, promoting cybersecurity awareness, and empowering individuals and communities to protect their digital freedoms, we can build a more resilient and secure society in the face of evolving threats.

THE FUTURE OF CONNECTIVITY: VIRTUAL WORLDS AND BEYOND

The future of connectivity is poised to redefine the very fabric of human existence, weaving together our physical and digital realities in ways that were once the realm of science fiction. This intricate tapestry of interconnectedness, woven with threads of innovation and possibility, promises to transform how we live, work, learn, and interact with the world. This chapter delves into the transformative potential of emerging technologies like virtual worlds and augmented reality, exploring the boundless opportunities they offer while addressing the growing concerns about cybersecurity in an increasingly interconnected world.

Imagine stepping into a virtual world, a vibrant digital realm where physical limitations dissolve and geographical boundaries cease. Here, you can collaborate with colleagues from across continents, explore fantastical landscapes, or even attend a concert with friends miles away, all while feeling a tangible sense of presence and shared experience. Virtual worlds offer a canvas for human interaction, education, and creative expression, expanding the horizons of human experience beyond the confines of our physical reality.

Augmented reality, on the other hand, overlays digital information onto the real world, enhancing our perception and interaction with our surroundings. Imagine walking down a city street and receiving personalized historical insights or directions through your AR glasses, or surgeons using AR overlays to visualize anatomical structures during complex procedures. This seamless blending of the physical and digital worlds can revolutionize industries, enhance productivity, and unlock new realms of creativity and innovation.

However, this exciting future of connectivity also presents new challenges, particularly cybersecurity. As our lives become increasingly intertwined with the digital world, the security of our data, privacy, and digital identities becomes paramount. The rise of cyberattacks, data breaches, and misinformation campaigns demands a proactive and multifaceted approach to cybersecurity, encompassing robust technological safeguards, comprehensive user education, and a collective commitment to responsible digital citizenship.

This chapter navigates the complex connectivity landscape, exploring emerging technologies' transformative potential while addressing cybersecurity's critical importance in an increasingly interconnected world. By understanding the opportunities and challenges ahead, we can harness the power of connectivity to shape a future where technology empowers individuals, fosters collaboration, and enriches the human experience.

VIRTUAL WORLDS: EXPANDING THE HORIZONS OF HUMAN EXPERIENCE

The rise of virtual worlds marks a profound shift in the human experience, offering a glimpse into a future where the boundaries between physical and digital realities blur. These immersive digital environments, where individuals can interact with each other and their surroundings in real-time, hold the promise of revolutionizing social interaction, learning, and creative expression. Imagine stepping into a vibrant virtual city where you can meet with colleagues from across the globe, attend a concert with friends who live thousands of miles away, or explore ancient ruins reconstructed in meticulous detail. These virtual spaces transcend the limitations of physical distance, creating new avenues for human connection and collaboration.

In education, virtual worlds offer immersive learning experiences that can transport students to distant lands, recreate historical events, or simulate scientific experiments in a safe and engaging environment. Imagine students exploring the Amazon rainforest, interacting with virtual indigenous communities, and learning about the ecosystem's delicate balance, all without leaving the classroom.

Virtual worlds provide a limitless canvas for creative expression for artists, musicians, and performers. Imagine attending a virtual concert where the boundaries of reality dissolve, and the audience becomes part of the performance, or exploring a virtual art gallery where sculptures defy gravity and paintings come to life.

The potential of virtual worlds extends far beyond entertainment and escapism. These digital realms offer new opportunities for collaboration, innovation, and social connection, fostering a sense of global community and empowering individuals to transcend the limitations of their physical surroundings. As we continue exploring the vast possibilities of virtual worlds, we embark on a journey that will redefine the human experience and reshape the future of social interaction, learning, and creative expression.

Imagine stepping into a virtual conference hall, a bustling hub of minds from every corner of the globe, gathered to exchange ideas and collaborate on groundbreaking projects. No longer bound by geographical constraints or the limitations of physical travel, participants can interact with each other as if they were indeed present, their avatars conversing and gesturing in a shared digital space. Imagine the vibrant discussions, the spontaneous brainstorming sessions, and the cross-cultural pollination of ideas that such a gathering could foster.

Alternatively, envision yourself immersed in a virtual museum, where the treasures of history are brought to life through interactive exhibits and immersive storytelling. Walk through the ancient ruins of Pompeii, explore the inner workings of a medieval castle, or examine the intricate brushstrokes of a Renaissance masterpiece, all from the comfort of your own home. These virtual experiences democratize access to education and culture, breaking down the barriers of distance and cost and allowing individuals from all walks of life to explore the world's rich heritage.

The potential of these immersive technologies extends far beyond entertainment and education. Virtual worlds can serve as platforms for global collaboration, enabling researchers, scientists, and innovators to work together on pressing challenges, regardless of their physical location. Imagine architects collaborating on a sustainable city design in a shared virtual space or medical professionals conducting a virtual surgery with colleagues from across continents.

The rise of virtual worlds and augmented reality heralds a new era of human interaction and collaboration, where the boundaries of physical reality blur, and the possibilities for connection and shared experiences are limitless. These technologies have the potential to democratize access to knowledge, foster cultural understanding, and promote global cooperation on an unprecedented scale, shaping a future where the collective intelligence of humanity is harnessed to address the challenges and opportunities of our interconnected world.

AUGMENTED REALITY: ENHANCING THE PHYSICAL WORLD

Augmented reality is poised to revolutionize our perception and interaction with the world. By seamlessly overlaying digital information onto our physical reality, AR opens up a realm of possibilities once confined to science fiction. Imagine surgeons guided by virtual overlays during complex procedures, students exploring ancient civilizations through interactive historical reconstructions, or architects visualizing their designs in three dimensions before laying a single brick.

The applications of AR are vast and transformative, spanning industries from healthcare and education to manufacturing and retail. AR can provide surgeons with real-time information during surgeries, enhancing precision and reducing risks. AR can bring textbooks to life in education, allowing students to interact with virtual models and simulations, fostering more profound understanding and engagement. AR can guide technicians through complex assembly processes in manufacturing, improving efficiency and reducing errors. In retail, AR can enable customers to visualize products in their homes before purchasing, enhancing the shopping experience and reducing buyer's remorse.

This seamless blending of the physical and digital realms has the potential to redefine our relationship with technology, making it an intuitive and integrated part of our daily lives. AR can enhance our perception of the world, providing us with access to information and experiences that were previously inaccessible. It can empower us with new knowledge, skills, and creative tools, expanding the horizons of human potential.

The future of augmented reality is bright, with advancements in hardware and software paving the way for even more immersive and transformative experiences. As AR technology evolves, we can expect to see more innovative applications emerge, blurring the lines between the physical and digital worlds and reshaping how we live, work, and interact with our surroundings.

Imagine a world where surgeons, peering through augmented reality visors, can visualize the intricate network of blood vessels beneath a patient's skin, guiding their instruments with unparalleled precision during complex procedures. Imagine students standing amidst the ruins of ancient Rome, not merely observing static remnants but experiencing the bustling city in its prime through interactive augmented reality tours that bring the past to life. These are glimpses of the transformative potential of augmented reality, a technology poised to revolutionize not only specialized fields but also the everyday experiences of individuals across the globe.

The possibilities are indeed boundless. AR overlays, seamlessly blending digital information with the real world, can enhance our perception, deepen our understanding, and expand our capabilities in ways we are only beginning to grasp. From education and healthcare to manufacturing and entertainment, AR is poised to reshape industries, redefine human-computer interaction, and ultimately transform how we live, work, and interact with the world around us.

This technology can potentially democratize knowledge and expertise, making complex information accessible to a broader audience. Imagine mechanics guided through intricate repairs by AR overlays highlighting components and providing step-by-step instructions, or tourists exploring foreign cities with AR applications that translate signs, provide historical context, and offer personalized recommendations.

Furthermore, AR can foster deeper engagement and emotional connection with the world. Imagine experiencing a symphony orchestra with AR visualizations that illuminate the interplay of instruments and enhance the emotional impact of the music, or visiting a natural history museum where AR overlays bring dinosaur skeletons to life, allowing visitors to witness their movements and behaviors in their ancient habitats.

The potential for AR to enhance human capabilities and transform our daily lives is immense, offering a future where the boundaries between the physical and digital worlds blur and a seamless blend of information, experience, and imagination enriches our perception of reality.

THE CYBERSECURITY IMPERATIVE: PROTECTING OUR DIGITAL FUTURE

While the tapestry of the hyper-connected world offers immense opportunities for progress and collaboration, it also casts a shadow of vulnerability. Our increasing reliance on technology and interconnectedness increases the risk of cyberattacks, data breaches, and the insidious manipulation of information. As we weave our lives ever more tightly into the digital fabric, we become increasingly susceptible to those who seek to exploit this dependence for malicious purposes.

The technologies that connect and empower us can also be turned against us, exposing our personal data, disrupting critical infrastructure, and eroding the trust underpinning our digital interactions. The hyper-connected world, therefore, demands a renewed and unwavering focus on cybersecurity awareness, education, and the development of robust security measures to protect individuals and communities in this digital age.

This necessitates a multi-pronged approach, encompassing technological advancements and a fundamental shift in our collective mindset. We must foster a culture of cybersecurity awareness, where individuals and organizations alike understand the importance of safeguarding their digital assets and recognize the evolving tactics of cyber adversaries. Education is crucial in empowering individuals with the knowledge and skills to navigate the digital landscape safely and responsibly, recognizing threats, protecting their privacy, and contributing to a more secure online environment.

Furthermore, developing robust security measures is paramount, encompassing technological safeguards policies and practices that prioritize data protection and privacy. This includes investing in research and development of cutting-edge security technologies, promoting international cooperation to combat cybercrime, and establishing ethical frameworks that guide the responsible use of technology.

In essence, the hyper-connected world presents a dual challenge: to harness technology's immense potential while safeguarding against its inherent vulnerabilities. By embracing cybersecurity awareness, education, and the development of robust security measures, we can navigate this complex landscape and ensure that the digital age remains a catalyst for progress, innovation, and a more secure future for all.

Securing our digital future requires a multifaceted approach, a concerted effort encompassing technological advancements, educational empowerment, and collaborative partnerships. This includes investing in research and development of advanced cybersecurity technologies, pushing the boundaries of innovation to stay ahead of evolving cyber threats. From sophisticated encryption algorithms and intrusion detection systems to AI-powered threat analysis and behavioral biometrics, these cutting-edge technologies will form the foundation of a resilient digital infrastructure.

Simultaneously, promoting education and awareness initiatives is crucial to empower individuals and communities to protect themselves online. This means fostering digital literacy, equipping individuals with the knowledge and skills to navigate the digital landscape safely and responsibly. It involves raising awareness about common cyber threats, such as phishing scams, social engineering tactics, and malware attacks, and providing practical guidance on recognizing and mitigating these risks.

Furthermore, fostering collaboration between governments, industry, and Academia is essential to creating a secure and resilient digital ecosystem. Governments play a vital role in establishing cybersecurity standards, promoting responsible innovation, and providing resources for education and research. With its expertise in technology development and deployment, industry can contribute to creating secure products and services while sharing threat intelligence and best practices. With its focus on research and education, Academia can drive innovation in cybersecurity technologies and contribute to developing a skilled cybersecurity workforce.

By investing in advanced technologies, empowering individuals through education, and fostering collaboration across sectors, we can build an interconnected, innovative, secure, and resilient digital world. This collective effort will pave the way for a future where technology empowers individuals, strengthens communities, and drives progress while safeguarding against the evolving threats of the digital age.

THE FUTURE OF HUMAN INTELLIGENCE IN A HYPER-CONNECTED WORLD

The hyper-connected world of the 21st century presents a complex and dynamic landscape, teeming with unprecedented opportunities and formidable challenges for human intelligence. Connecting with individuals across the globe instantaneously, accessing a vast ocean of information with a few keystrokes, and immersing our-selves in virtual realities that blur the lines between the physical and digital realms offers immense potential for collaboration, innovation, and societal advancement. However, this hyper-connectivity also casts a shadow, raising concerns about the erosion of privacy, the manipulation of information, and the potential for cognitive overload.

The challenges of censorship loom large in this digital age, where governments and powerful entities seeking to control the narrative and suppress dissent can cur-tail the free flow of information. The technologies that connect us can also monitor our online activities, restrict access to knowledge, and manipulate our perceptions. The potential impact on human cognition is another area of concern. The constant bombardment of information, the distractions of the digital world, and the increasing reliance on artificial intelligence raise questions about our attention spans, critical thinking skills, and the very nature of human thought.

Furthermore, the hyper-connected world presents an ever-expanding attack sur-face for cyber threats. The more we rely on technology for communication, com-merce, and critical infrastructure, the more vulnerable we become to cyberattacks, data breaches, and the disruption of essential services. These threats demand a pro-active and multifaceted approach to cybersecurity, encompassing technological safe-guards, education, awareness, and international cooperation.

We must navigate these challenges with wisdom and foresight to ensure that the hyper-connected world catalyzes a more informed, equitable, and secure future for all. This requires a commitment to digital literacy, critical thinking, and technol-ogy's ethical development and deployment of technology. It necessitates a global dia-logue on cybersecurity, privacy, and the responsible use of artificial intelligence. By embracing the opportunities and addressing the challenges of the hyper-connected world, we can harness its transformative power to foster a future where technology empowers individuals, strengthens communities, and promotes a more just and equi-table global society.

Navigating the complexities of our hyper-connected world demands a holistic approach that thoughtfully considers the ethical implications of emerging technolo-gies, actively promotes digital literacy and critical thinking skills, and fosters a cul-ture of responsible innovation and collaboration. This requires a shift in perspective, moving beyond the mere adoption of technology to a deeper understanding of its potential impact on individuals, communities, and the very fabric of society.

Ethical considerations must be at the forefront of technological development and deployment. We must critically examine the potential consequences of new tech-nologies, ensuring they align with human values, promote fairness and inclusivity, and safeguard against unintended biases or discriminatory outcomes. This involves fostering open dialogue among technologists, ethicists, policymakers, and the public to ensure that technological advancements serve the common good and contribute to a more just and equitable society.

Promoting digital literacy and critical thinking skills is paramount in empower-ing individuals to navigate the digital landscape effectively and responsibly. This involves equipping individuals with the ability to discern credible information from misinformation, evaluate online content critically, and engage in thoughtful online discourse that fosters understanding and avoids the pitfalls of echo chambers and filter bubbles.

Furthermore, fostering a responsible innovation and collaboration culture is essential. This involves encouraging the development of technologies that serve human needs and promote social well-being while promoting collaboration across disciplines and sectors to ensure that technological advancements are guided by ethical principles and contribute to a sustainable future.

By embracing the opportunities and proactively addressing the challenges of the hyper-connected world, we can harness the transformative power of technology to enhance human intelligence, foster social progress, and create a more equitable and sustainable future for all. This requires a collective commitment to shaping a digital world that reflects our values, empowers individuals, and strengthens the bonds of our shared humanity.

THE RISE OF CULTURAL RESILIENCY

While technology has undeniably woven a web of global interconnectedness, linking individuals and communities in unprecedented ways, it has also inadvertently sparked a counter-movement: a resurgence of cultural resiliency against the perceived dominance of technological intelligence. This resiliency, a testament to the enduring human spirit and the desire to safeguard cultural identity, manifests in diverse forms, from revitalizing traditional art forms and preserving indigenous languages to active resistance against censorship and the erosion of privacy.

In a world increasingly saturated with digital technology, where algorithms shape our experiences and artificial intelligence encroaches upon human domains, a growing awareness of the need to preserve and celebrate cultural heritage has emerged. Communities are rediscovering the richness of their traditions, revitalizing ancient art forms, and reclaiming their cultural narratives in the face of homogenizing global trends. The preservation of indigenous languages, often threatened by the dominance of global languages, has become a rallying cry for cultural identity and self-determination.

Moreover, the erosion of privacy and the rise of surveillance technologies have sparked a wave of resistance, with individuals and communities demanding greater control over their digital footprints and personal data. The fight against censorship, particularly in the digital realm, has intensified as individuals and organizations strive to protect freedom of expression and access to unfiltered information.

This resurgence of cultural resiliency underscores the complex and dynamic relationship between technology and culture. While technology offers immense potential for cultural exchange and preservation, it also carries the risk of homogenization, manipulation, and control. The challenge lies in harnessing the benefits of technology while safeguarding cultural diversity and individual autonomy.

In an increasingly digital world, where algorithms and artificial intelligence subtly shape our experiences and influence our perceptions, individuals and communities seek to reclaim control over their cultural narratives and digital identities. The attempts by governments and corporations to restrict access to information, limit freedom of expression, and manipulate online narratives have fueled this cultural resistance, sparking a backlash against the homogenizing forces of technological globalization.

This cultural resiliency is evident in the rise of decentralized platforms that challenge the dominance of centralized social media giants, offering alternative spaces for expression and community building that prioritize user autonomy and data privacy. It is also apparent in the development of privacy-enhancing technologies that protect individuals from surveillance and data exploitation, empowering them to navigate the digital landscape with greater control over their personal information. Moreover, there is a growing awareness of the importance of digital literacy and critical thinking in navigating the complex digital landscape, enabling individuals to discern fact from fiction, evaluate information sources, and engage in thoughtful online discourse.

This resurgence of cultural resiliency is a testament to the enduring human desire for autonomy, self-expression, and the preservation of cultural heritage in the face of technological dominance. It raises important questions about the future of human intelligence and the role of technology in shaping our societies. Will technology continue to dominate and homogenize cultural expressions, eroding the rich tapestry of human experience? Or will we find ways to harness its power while preserving our diverse cultural heritage, fostering intercultural understanding, and empowering individuals and communities?

The answer lies in our collective ability to critically examine the impact of technology on our lives, to engage in open and honest dialogue about its ethical implications, and to actively shape a future where technology empowers individuals and communities rather than diminishing their autonomy and cultural identity. By promoting digital literacy, fostering critical thinking, and advocating for responsible technology governance, we can ensure that the digital age catalyzes cultural flourishing and a more equitable and interconnected world.

NAVIGATING THE COMPLEXITIES OF TECHNOLOGICAL INFLUENCE

The relationship between technology and culture is a complex and multifaceted tapestry woven with both opportunity and challenge threads. Technology can be a powerful tool for cultural exchange and preservation, enabling sharing stories, traditions, and artistic expressions across vast distances; it also carries the potential for manipulation, control, and the homogenization of cultural identities.

The challenge lies in finding a delicate balance between harnessing the benefits of technology while safeguarding cultural diversity and individual autonomy. This requires a nuanced approach that recognizes the importance of cultural resiliency – the ability of cultures to adapt and thrive in the face of technological change – while promoting responsible technology governance and ethical innovation.

Fostering open dialogue and critical engagement with technology is essential. By encouraging discussions about the ethical implications of new technologies, we can ensure that their development and deployment align with human values and cultural diversity. Promoting digital literacy empowers individuals and communities to navigate the digital landscape critically and consciously, making informed choices about how they engage with technology and shape its impact on their cultural identities.

Furthermore, empowering individuals and communities to shape their technological landscape is crucial. This involves promoting participatory design processes, ensuring diverse representation in technology development, and supporting initiatives that leverage technology to preserve and revitalize cultural heritage.

We can chart a course through the complex and often turbulent relationship between technology and culture by embracing these principles – open dialogue, digital literacy, and community empowerment. This will help ensure that the digital age, rather than being a homogenizing force, becomes a catalyst for cultural flourishing, fostering a world where diversity is celebrated and interconnectedness strengthens our shared human experience.

Imagine a world where technology empowers diverse cultural expressions, providing platforms for marginalized voices to be heard, for ancient traditions to be preserved and shared, and for artistic creativity to blossom in new and vibrant forms. Imagine a world where digital tools facilitate intercultural understanding, breaking down barriers of language and geography, enabling the exchange of ideas and perspectives, and fostering empathy and collaboration across cultures. Imagine a world where technology strengthens the bonds of our shared human experience, connecting individuals and communities across continents, celebrating our common humanity, and inspiring collective action toward a more just and sustainable future.

We must strive for this vision – a world where technology serves not as a master but as a tool, empowering individuals, strengthening communities, and fostering a global culture that celebrates diversity, promotes understanding, and embraces our shared human journey. By embracing the principles of open dialogue, digital literacy, and community empowerment, we can navigate the complex relationship between technology and culture, ensuring that the digital age leaves a legacy of cultural flourishing and a more equitable and interconnected world.

THE FIREWALL OF RESISTANCE: CYBERSECURITY, CULTURAL RESILIENCY, AND THE BATTLE FOR SOCIETAL INTELLIGENCE

Cybersecurity has undergone a profound transformation in an era defined by the pervasive reach of digital connectivity and the ubiquitous presence of technology. No longer confined to technical safeguards and firewalls, it has become deeply intertwined with the resilience of cultures and the collective intelligence of societies. This chapter delves into the intricate interplay between these forces, particularly within societies grappling with the suffocating grip of censorship and the overbearing control of authorities. It examines how the rise of cultural resiliency, fueled by the imperative to protect digital freedoms and maintain unfettered access to information, has emerged as a formidable challenge to those who seek to manipulate the flow of knowledge and suppress dissent.

This cultural resistance, born out of a deep-seated yearning for autonomy and intellectual freedom, manifests in many ways. It can be witnessed in the clandestine networks of individuals utilizing encryption and anonymity tools to circumvent censorship barriers, in the collective defiance of online communities challenging government-imposed restrictions, and in the unwavering determination of journalists and activists utilizing secure platforms to disseminate information and expose abuses of power.

This chapter will explore the dynamic interplay between cybersecurity, cultural resilience, and societal intelligence, examining the strategies employed by those seeking to control information and those fighting to liberate it. It will analyze the impact of censorship on the collective intelligence of societies and how the struggle for digital freedom is shaping the future of human expression, knowledge sharing, and social progress in the digital age.

CYBERSECURITY: A PILLAR OF CULTURAL RESISTANCE

In societies stifled by censorship and pervasive surveillance, cybersecurity transcends its traditional role of protecting data and systems. It evolves into a powerful instrument of cultural resistance, a digital shield against the encroachment of controlling authorities. The ability to circumvent censorship, safeguard digital identities, and secure communication channels becomes paramount, empowering individuals and communities to challenge oppressive regimes and maintain access to unfiltered information.

Encryption, once the domain of tech-savvy specialists, transforms into a sword of defiance, rendering communications impervious to prying eyes. Virtual Private Networks (VPNs) become secret tunnels, allowing citizens to bypass government-imposed barriers and access a world of information beyond the digital walls erected by their oppressors. These tools, once considered mere technical safeguards, now symbolize the fight for freedom of thought and expression.

Utilizing these cybersecurity measures becomes a subtle yet powerful rebellion, a declaration of independence in the digital sphere. Citizens equipped with these tools can access blocked websites, engage in uncensored discussions on social media platforms, and tap into the lifeblood of independent news sources, fostering a more informed and empowered populace.

This digital resistance, however, extends beyond individual empowerment. It strengthens social bonds by providing secure channels for like-minded individuals to connect, organize, and mobilize collective action. The shared struggle against censorship fosters a sense of solidarity, creating a virtual community united in its pursuit of truth and freedom.

In essence, cybersecurity becomes a cornerstone of cultural resiliency, a vital force in preserving intellectual freedom and challenging the oppressive grip of censorship. It enables citizens to reclaim control over their digital lives, fostering a more informed, connected, and ultimately resilient society capable of resisting the erosion of fundamental rights.

CULTURAL RESILIENCY: A CHALLENGE TO CONTROLLING AUTHORITIES

The rise of cultural resiliency, fueled by the imperative to protect digital freedoms, throws a formidable wrench into the gears of controlling authorities. Their attempts to maintain a tight grip on the narrative and suppress dissenting voices are increasingly undermined by the ability of individuals and communities to circumvent censorship, organize online, and disseminate information with remarkable agility. This dynamic represents a fascinating paradox of the digital age: the very technologies intended to consolidate power and control have become instruments of empowerment and resistance in the hands of determined citizens.

Once envisioned by some as a tool for surveillance and propaganda, the internet has evolved into a fertile ground for the flourishing of counter-narratives and the mobilization of dissent. Despite being subject to manipulation and censorship, social media platforms serve as vital channels for sharing information, organizing protests, and galvanizing public opinion. Encryption technologies and anonymization tools empower individuals to bypass censorship firewalls and access unfiltered information, while decentralized platforms offer alternative spaces for expression and collaboration beyond the reach of centralized control.

This rise of cultural resiliency poses a significant challenge to authoritarian regimes and those seeking to maintain control over the flow of information. The ability of individuals and communities to circumvent censorship, organize online, and disseminate information with remarkable speed and agility undermines the foundations of their power. The internet, once viewed as a threat to their authority, has become a catalyst for empowerment, enabling citizens to challenge the status quo, demand accountability, and assert their right to freedom of expression.

This dynamic highlights the complex interplay between technology, culture, and power in the digital age. The struggle for control over the narrative is no longer confined to traditional media and physical spaces; it has spilled over into the virtual realm, where the battle lines are drawn in the bits and bytes of cyberspace. The rise of cultural resiliency demonstrates that the human desire for freedom and self-expression cannot be easily contained, even in the face of sophisticated censorship and surveillance technologies.

In a twist of irony, the technologies designed to facilitate surveillance and control have been repurposed as empowerment tools in the hands of resilient citizens. Social media platforms, intended to monitor and manipulate online discourse, have been transformed into vital arteries for the lifeblood of resistance, pulsating with unfiltered information and defiant voices. Despite being subject to censorship and manipulation, these platforms serve as clandestine channels for sharing uncensored news, coordinating protests, and mobilizing public opinion, effectively challenging the authorities' attempts to control the flow of information.

This dynamic creates a constant tug-of-war, a digital battlefield where the forces of control clash with the unwavering spirit of a connected and determined populace. The censors erect firewalls, while the citizens develop virtual private networks to bypass them. The authorities attempt to manipulate narratives while the people share unfiltered accounts and expose the truth. The struggle is ongoing, a testament to the enduring human desire for freedom of expression and the power of technology to amplify marginalized voices.

This digital battlefield is not merely a contest of technical prowess but a clash of ideologies, a struggle between the forces of control and the unwavering spirit of a connected and determined populace. The outcome of this struggle will shape the future of our societies, determining whether technology will be used to empower or enslave, enlighten or deceive.

RESILIENT SOCIETIES: POWERHOUSES OF INTELLIGENCE

Resilient societies champion digital freedoms and the unimpeded flow of information, and become fertile breeding grounds for intellectual curiosity and innovation. Like a vibrant ecosystem teeming with diverse life forms, these societies thrive on the open exchange of ideas, the clash of perspectives, and the freedom to challenge long-held assumptions. This intellectual freedom is not simply the absence of censorship but a proactive commitment to fostering an environment where curiosity is nurtured, dissent is valued, and the pursuit of knowledge knows no bounds.

This creates a self-sustaining cycle of intellectual growth and societal advancement. Open access to information fuels critical thinking, empowering individuals to question, analyze, and synthesize knowledge from diverse sources. This, in turn, fosters a spirit of inquiry, a willingness to venture beyond the confines of conventional wisdom and explore uncharted territories of thought. The result is a society constantly pushing the boundaries of knowledge, generating novel solutions, and driving progress across all fields of human endeavor.

In these resilient societies, innovation is not merely a byproduct of intellectual freedom; it is woven into the very fabric of their cultural DNA. The freedom to experiment, challenge the status quo, and learn from failures creates a fertile ground for creativity and ingenuity. This spirit of innovation permeates all levels of society, from bustling research labs and vibrant artistic communities to the everyday interactions of individuals empowered to think critically and challenge conventional wisdom.

The benefits of such an environment extend far beyond technological advancements or scientific breakthroughs. With their commitment to open knowledge and intellectual freedom, resilient societies are better equipped to address complex challenges, adapt to changing circumstances, and build a more just and equitable future for all. They are societies where the pursuit of knowledge is not merely a means to an end but a fundamental expression of human potential, a testament to our capacity to learn, grow, and shape a better world for ourselves and future generations.

Conversely, societies that operate under a shroud of censorship and suppression create an intellectual vacuum where the free exchange of ideas is stifled, and the pursuit of knowledge is constrained. This suppression of information limits access to diverse perspectives, breeds conformity, and discourages the critical thinking essential for societal progress. Innovation withers in such an environment, as individuals hesitate to challenge the status quo or explore ideas that deviate from the accepted narrative. Fear of reprisal, social ostracization, or even legal consequences casts a chilling effect on intellectual discourse, discouraging the open exchange of ideas that fuels creativity and drives progress.

The censorship of ideas and the suppression of dissenting voices ultimately hinder a society's ability to adapt, evolve, and progress. Such societies become stagnant, trapped in echo chambers where the same ideas and perspectives are recycled, reinforcing existing biases and hindering the emergence of new solutions to societal challenges. The lack of exposure to diverse viewpoints and the suppression of critical inquiry create a breeding ground for intellectual stagnation, where dogma reigns supreme, and the potential for societal advancement is severely limited.

In contrast, societies that embrace freedom of expression and cultivate open dialogue foster a vibrant intellectual landscape where innovation flourishes. The collision of diverse perspectives, the challenge of assumptions, and the freedom to explore unconventional ideas create a fertile ground for creativity, problem-solving, and progress. Such societies are better equipped to adapt to change, embrace new challenges, and thrive in an increasingly complex and interconnected world.

THE ROLE OF TECHNOLOGY IN FOSTERING RESILIENCY

In its inherently dualistic nature, technology acts as both a sword and a shield in the digital age's struggle for control and freedom. It is a double-edged sword capable of empowering, oppressing, connecting and dividing, illuminating and obscuring. On one hand, technology provides powerful tools for surveillance and censorship, enabling authorities to monitor online activities, restrict access to information, and manipulate narratives. Governments and corporations can deploy sophisticated technologies to track individuals' movements, intercept communications, and censor dissenting voices, creating an environment where freedom of expression and access to information are curtailed.

On the other hand, technology also empowers individuals and communities to resist these controlling forces, providing the means to circumvent censorship, protect privacy, and foster open communication. Encryption tools shield sensitive information from prying eyes, anonymization networks allow individuals to express themselves without fear of reprisal, and decentralized platforms create spaces for uncensored dialogue and collaboration.

This duality of technology creates a constant tension between control and freedom, a dynamic interplay that shapes the digital landscape and influences the evolution of human society. The same technologies used to monitor and suppress dissent can also be harnessed to expose wrongdoing, organize resistance, and promote transparency. The outcome of this struggle depends on the choices we make, the values we prioritize, and how we choose to wield the powerful tools of the digital age.

This inherent duality demands a critical and nuanced approach to technology that recognizes its potential for empowerment and its capacity for oppression. We must safeguard digital freedoms, promote ethical innovation, and ensure that technology catalyzes a more just, equitable, and interconnected world.

The rise of privacy-enhancing technologies has become a crucial shield in the defence of individual autonomy in the digital age. Encryption, like a digital lock, scrambles information into an unreadable format, ensuring that only authorized individuals with the corresponding key can access its true meaning. This safeguards sensitive data, from personal communications to financial transactions, from prying eyes and malicious actors seeking to exploit vulnerabilities. Anonymization tools further enhance privacy by masking the identities of individuals, allowing them to navigate the digital landscape without fear of surveillance or tracking. This empowers individuals to express themselves freely, engage in open dialogue, and explore diverse perspectives without the chilling effect of constant monitoring.

Decentralized platforms have emerged as a powerful counterforce to the dominance of monolithic social media giants. Operating independently of centralized control, these platforms offer alternative spaces for expression, collaboration, and information sharing, fostering a more democratic and resilient digital landscape. By distributing control among a network of users, decentralized platforms reduce the risk of censorship, data manipulation, and the abuse of power that can plague centralized systems. This fosters a more diverse and vibrant online ecosystem where individuals have greater control over their data and digital identities.

Secure communication channels, fortified by robust encryption and authentication protocols, act as lifelines for the free flow of information, even in the face of censorship and surveillance that seek to suppress it. These digital fortresses, built on complex algorithms and cryptographic keys, allow individuals to communicate without fear of eavesdropping, to share ideas without the threat of reprisal, and to access information that would otherwise be hidden behind digital walls.

This dynamic interplay between technology's potential for control and its capacity for liberation has created a crucial battleground in the ongoing struggle for digital freedoms. It is a digital tug-of-war, where governments and corporations seek to exert control over the flow of information while individuals and communities leverage technology to resist these constraints and reclaim their right to access, share, and express themselves freely. The rise of cultural resiliency, fueled by the imperative to protect these freedoms and maintain open access to information, has become a formidable challenge to controlling authorities. It is a testament to the human spirit's refusal to be silenced, a digital David versus Goliath where the slingshot of encryption and the stones of secure communication are used to topple the giants of censorship and surveillance.

This struggle is not merely technological; it is a battle for the very soul of the digital age. Will it be an era of open communication, shared knowledge, control, manipulation, and restricted access? The answer lies in the hands of individuals, communities, and organizations who recognize the power of technology to both liberate and oppress. By embracing the tools of secure communication, advocating for digital rights, and fostering a culture of cybersecurity awareness, we can tilt the balance toward a future where the digital world truly empowers individuals and fosters a more informed, interconnected, and equitable society.

Resilient societies that embrace the free flow of information and a steadfast commitment to open knowledge become veritable powerhouses of intelligence, driving innovation, progress, and social change. Like a vibrant ecosystem teeming with diverse life forms, these societies thrive on the open exchange of ideas, the clash of perspectives, and the relentless questioning of conventional wisdom. This intellectual ferment fuels a culture of critical inquiry, where individuals are empowered to challenge assumptions, explore new possibilities, and contribute to the collective pursuit of knowledge.

In these societies, technology serves as an enabler, amplifying the power of human connection and facilitating the dissemination of information across geographical and cultural boundaries. The internet, social media platforms, and other digital tools have become instruments of empowerment, allowing citizens to access diverse sources of information, engage in open dialogue, and hold those in power accountable.

In stark contrast, societies that succumb to the allure of censorship and control create an intellectual vacuum where creativity is stifled, innovation is hindered, and progress stagnates. Like a barren landscape devoid of nutrients, these societies lack the essential ingredients for growth and development. The suppression of dissenting voices, the restriction of information, and the fear of challenging the status quo create an atmosphere of intellectual conformity, where the enforcement of dogma replaces the pursuit of knowledge.

This suppression of intellectual freedom hinders scientific and technological advancement and undermines social progress and cultural development. The lack of open dialogue and critical inquiry leads to a stagnation of ideas, a perpetuation of outdated beliefs, and an inability to adapt to the challenges of a rapidly changing world. Ultimately, societies that choose the path of censorship and control diminish their potential, forfeiting the opportunity to harness their citizens' collective intelligence and creativity.

As the digital landscape continues its relentless and rapid evolution, the intricate interplay between cybersecurity, cultural resiliency, and societal intelligence will remain a defining characteristic of the 21st century and beyond. This dynamic dance, where technological advancements intertwine with cultural values and societal adaptations, will shape the trajectory of human progress and determine the fate of nations in the digital age.

The societies that thrive in this dynamic environment will be those that not only embrace the empowering potential of technology but also remain vigilant in safeguarding digital freedoms and fostering a culture of open knowledge and critical engagement. They will be the societies that recognize the transformative power of technology while acknowledging its potential for misuse and manipulation.

These societies will champion cybersecurity awareness, educating citizens about the evolving landscape of cyber threats and empowering them with the knowledge and skills to navigate the digital world safely and responsibly. They will invest in robust cybersecurity infrastructure, protecting critical systems and sensitive data from malicious actors while ensuring the privacy and security of their citizens.

Furthermore, these societies will cultivate a culture of open knowledge and critical engagement, recognizing that the free flow of information, the diversity of perspectives, and the ability to challenge conventional thinking are essential for progress and innovation. They will resist the temptation to censor ideas, suppress dissent, or manipulate narratives, understanding that such actions ultimately stifle creativity, hinder progress, and erode trust.

In essence, the societies that thrive in the digital age will embrace technology as a tool for empowerment, a catalyst for progress, and a bridge to connect individuals and communities across geographical and cultural divides. They will be the societies that foster a culture of cybersecurity awareness, digital literacy, and critical engagement, ensuring that the digital revolution serves as a force for good, propelling humanity toward a more informed, interconnected, and equitable future.

While seemingly paradoxical, technologies initially designed for surveillance and control can, in the hands of a resilient populace, be repurposed to foster and strengthen resistance. Web ghosting, obscuring one's digital footprint and circumventing surveillance, becomes crucial for those seeking to exercise their digital freedoms in restrictive environments. By utilizing anonymization tools, encrypted communication channels, and decentralized platforms, individuals can reclaim control over their digital identities and navigate the online world with greater autonomy.

Surveillance technologies intended to monitor and track individuals can be subverted and used to expose abuses of power and hold authorities accountable. Citizen journalism, enabled by readily available recording devices and online platforms, allows individuals to document and disseminate evidence of human rights violations, corruption, and social injustices, often challenging official narratives and galvanizing public support for change.

Moreover, the awareness of being surveilled can paradoxically foster a sense of solidarity and collective resistance. When individuals know their actions are being monitored, they may be more inclined to seek out secure communication channels, utilize privacy-enhancing technologies, and engage in collective action to protect their digital freedoms.

Resisting surveillance and censorship can become a form of cultural expression, a symbol of defiance against attempts to control information and suppress dissent. This cultural resiliency, fueled by the desire for autonomy and self-expression, can inspire innovation and creativity, leading to the development of new tools and strategies to circumvent restrictions and reclaim digital freedoms.

The relationship between surveillance technologies and cultural resiliency is complex and dynamic. While surveillance can suppress dissent and control populations, it can also inadvertently foster resistance, innovation, and a renewed commitment to digital freedoms. The societies that thrive in the digital age will recognize this dynamic and actively shape a technological landscape that empowers individuals, protects privacy, and fosters a culture of open knowledge and critical engagement.

Technologies like license plate bright covers play a multifaceted role in fostering cultural resiliency in the digital age. By providing a layer of anonymity and control over personal information, these technologies empower individuals to resist surveillance, challenge controlling authorities, and assert their digital freedoms.

When activated, license plate bright covers obscure the license plate number from automated recognition systems, such as those used for traffic monitoring, toll collection, and law enforcement surveillance. This technology allows individuals to move through public spaces anonymously, reducing the digital footprint that is increasingly used to track and profile individuals.

In societies where surveillance is pervasive, the ability to control the visibility of one's license plate number becomes an act of resistance, a way to reclaim a measure of privacy and autonomy in the face of encroaching technological control. It allows individuals to challenge the assumption that their movements and activities should be constantly monitored and recorded.

Moreover, license plate smart covers can be a powerful tool for collective action and social protest. By obscuring their license plate numbers, individuals participating in demonstrations or rallies can reduce the risk of identification and retaliation by authorities. This can embolden citizens to exercise their right to freedom of assembly and expression without fear of reprisal.

Using such technologies also highlights the growing awareness of the importance of digital privacy and the right to control one's personal information. It signals a shift in societal attitudes toward surveillance, with individuals and communities increasingly demanding greater transparency and accountability from those who collect and use their data.

Technologies like license plate bright covers symbolize cultural resiliency, empowering individuals to resist surveillance, challenge controlling authorities, and assert their digital freedoms. They represent a proactive step toward reclaiming control over one's digital identity and fostering a more equitable and privacy-conscious society.

While a supportive and nurturing family environment can provide a strong foundation for personal growth and resilience, the unfortunate reality is that not all families embody these ideals. In some cases, family dynamics can be a source of stress, conflict, and even trauma, profoundly impacting an individual's development and their ability to cope with adversity.

Dysfunctional family relationships, characterized by patterns of neglect, emotional abuse, or inconsistent parenting, can leave deep-seated scars that shape an individual's sense of self-worth, their ability to form healthy relationships, and their capacity for resilience. Children who grow up in such environments may develop coping mechanisms that, while serving to protect them in the short term, can hinder their ability to navigate challenges and setbacks in the long run.

However, despite adverse family experiences, the human spirit has a remarkable capacity for resilience. Individuals who have endured complicated family dynamics often develop a profound inner strength, a fierce independence, and a deep empathy for others who have faced similar struggles. Though painful, these experiences can forge a resilience born of adversity and determination to overcome challenges and create a better life for themselves and others.

The role of bad families and relationships in fostering resilience is complex and multifaceted. While these experiences can undoubtedly create obstacles and challenges, they can also catalyze personal growth and transformation. Individuals who have navigated complex family dynamics often develop a profound understanding of human behavior, a heightened awareness of their strengths and vulnerabilities, and a deep appreciation for the importance of healthy relationships.

These experiences can also fuel a desire to break free from destructive patterns, create healthier and more fulfilling relationships, and build a better future for themselves and their families. The resilience forged in adversity can be a powerful force for positive change, empowering individuals to overcome challenges, advocate for themselves and others, and create a more just and compassionate world.

In the intricate dance between societal pressures, individual freedoms, and technological advancements, cultural resiliency emerges as a powerful force, shaping the trajectory of human societies and their capacity for progress and innovation. This chapter explores the complex interplay between these forces, highlighting the role of families and societal pressures in fostering resilience, particularly in the face of controls and censorship.

The imposition of censorship by families, societal institutions, or governing bodies creates a stifling environment that can hinder intellectual curiosity, limit freedom of expression, and foster a culture of fear and mistrust. However, this suppression can also catalyze cultural resiliency, igniting a flame of resistance and driving individuals and communities to seek alternative avenues for expression, knowledge sharing, and preserving cultural identity.

In its inherently dualistic nature, technology plays a complex role in this dynamic. On one hand, it can be a control and surveillance tool, enabling censorship enforcement and the suppression of dissent. On the other hand, technology also empowers individuals and communities to circumvent restrictions, access information, and connect with like-minded individuals across geographical boundaries.

The rise of decentralized platforms, encryption technologies, and anonymization tools allows individuals to challenge censorship, protect their digital privacy, and engage in open dialogue. Despite attempts to control and restrict the internet, it has become a fertile ground for cultural expression, social mobilization, and the dissemination of alternative narratives.

The interplay between societal pressures, cultural resiliency, and technological advancements is shaping the future of human societies. The societies that thrive in this dynamic environment will embrace the empowering potential of technology while remaining vigilant in safeguarding digital freedoms and fostering a culture of open knowledge and critical engagement.

As we navigate the complexities of the 21st century and beyond, the lessons learned from the interplay between censorship, cultural resiliency, and technological innovation will be crucial in shaping a future where human intelligence flourishes, societies progress, and the human spirit remains resilient in the face of adversity.

3 The Evolution of Societal Intercommunal Intelligence

From Ancient Messengers to the Hyper-Connected Digital Age

In the annals of human history, the evolution of societal intercommunal intelligence has been a long and winding journey, shaped by the interplay of communication technologies, social structures, and the ever-present challenges of security and trust. This chapter embarks on a captivating exploration of this journey, tracing intelligence development from the era of slow travel and handwritten letters to the hyper-connected digital age, where information flows at the speed of light and the boundaries between community's blurs.

In the nascent stages of human civilization, when the whispers of knowledge were carried on the winds of human messengers and etched onto fragile parchments, the pace of intercommunal intelligence was a slow and deliberate dance. Precious and fragile information traversed vast distances along trade routes' arteries and diplomatic channels' delicate threads. Envoys entrusted with secrets of state, merchants bearing tales of distant lands, and spies cloaked in shadows all played their part in this intricate web of intelligence gathering.

The challenges were immense. Spies relied on their cunning and guile to infiltrate foreign courts and glean whispers of hidden agendas. Scribes meticulously encoded messages in intricate ciphers, hoping to safeguard sensitive information from prying eyes. Leaders and advisors grappled with deciphering these coded missives and interpreting reports often clouded by bias, misinformation, or outright deception. The pursuit of knowledge in this era was a high-stakes game of whispers and shadows, where trust was a rare commodity, and the truth often remained elusive.

The invention of the printing press irrevocably altered the course of human communication, accelerating the dissemination of knowledge and ushering in a new era of intellectual exchange. Where once the written word had been the domain of the elite few, painstakingly copied by hand, books, pamphlets, and newspapers could now be produced relatively quickly and disseminated widely. Ideas, once confined to cloistered monasteries and scholarly circles, now flowed freely through the veins of society, challenging traditional hierarchies, sparking

DOI: 10.1201/9781003641506-3

fervent debates, and fueling social and political movements that reshaped the world.

However, this newfound freedom of information also brought new challenges. The technologies that democratized knowledge also provided tools for those in power to manipulate and control the flow of information. Censorship, propaganda, and the suppression of dissenting voices became sophisticated tactics employed to maintain control and stifle challenges to authority. The battle for intellectual freedom was far from over; it had simply entered a new and more complex arena.

The advent of the telegraph, telephone, and radio marked another seismic shift in the communication landscape, ushering in an era of instantaneity. The world shrunk as news and information traversed continents in seconds, further accelerating the pace of intercommunal intelligence. The ability to communicate across vast distances in real-time fostered a sense of global interconnectedness, facilitating trade, diplomacy, and cultural exchange on an unprecedented scale.

However, this rapid acceleration of communication also brought new challenges to security and privacy. The interception and manipulation of messages, once a laborious task requiring physical access, became a matter of technological prowess. The challenges of securing these communication channels and protecting sensitive information from prying eyes grew exponentially, demanding new strategies and technologies to safeguard the integrity and confidentiality of communications in an increasingly interconnected world.

Today, in the hyper-connected digital age, the flow of information has become a torrent, overwhelming traditional boundaries and creating a global village where communities are intertwined in unprecedented ways. The internet, social media, and mobile devices have revolutionized the way we communicate, learn, and interact with each other, creating both opportunities and challenges for societal intelligence.

The ease of information sharing and the rapid dissemination of knowledge have fostered collaboration, innovation, and cultural exchange on a global scale. However, the challenges of misinformation, cyberattacks, and the erosion of privacy have also become more acute. The very technologies that connect us also expose us to new vulnerabilities, making it increasingly difficult to understand information's origins, intentions, and trustworthiness.

Looking ahead, the future of intercommunal intelligence will be inextricably woven with the rise of virtual worlds, the pervasive influence of artificial intelligence, and the ever-deepening integration of technology into the fabric of our lives. In this dynamic and rapidly evolving landscape, the challenges of Censorship, control, and the preservation of human autonomy will become even more pronounced, demanding careful consideration and proactive measures to safeguard the integrity of information and the freedom of thought.

The emergence of virtual worlds, immersive digital environments where individuals can interact and collaborate, presents exciting opportunities and potential pitfalls for intercommunal intelligence. While these virtual spaces can foster connection, creativity, and knowledge exchange across geographical boundaries, they also raise concerns about the potential for manipulation, surveillance, and the erosion of privacy.

The increasing reliance on artificial intelligence for information processing and decision-making adds another layer of complexity. While AI can enhance efficiency and provide valuable insights, it raises concerns about algorithmic bias, the potential for misuse, and the erosion of human control over critical systems.

In this technologically mediated landscape, Censorship and control challenges loom. As governments and corporations seek to exert influence over the flow of information and shape online narratives, preserving digital freedoms and accessing unfiltered information become paramount.

The nature of intercommunal intelligence, built on the trust and reliability of information shared between communities, is vulnerable to compromise in this environment. Deepfakes, sophisticated AI-generated forgeries, can blur the lines between reality and fabrication, making it increasingly difficult to discern truth from falsehood. Cyberattacks aimed at disrupting critical infrastructure or manipulating data can sow discord, erode trust, and undermine the foundations of societal cooperation.

As we navigate this complex and ever-evolving digital landscape, preserving human autonomy and the ability to think critically and independently becomes even more crucial. The future of intercommunal intelligence hinges on our ability to harness the empowering potential of technology while remaining vigilant in safeguarding digital freedoms, fostering a culture of open knowledge, and promoting media literacy and critical engagement with information.

This chapter delves into these complex issues, tracing the evolution of societal intercommunal intelligence from the early days of slow travel and handwritten letters to the hyper-connected digital age and beyond. By understanding the historical context, the current challenges, and the potential future trajectories, we can gain valuable insights into the dynamics of knowledge sharing, the importance of cybersecurity, and the enduring human quest for connection, understanding, and progress in an ever-changing world.

HISTORY

Throughout history, the pursuit of knowledge and the ability to gather, analyze, and utilize information have been essential for the survival and advancement of societies. This insatiable quest for intelligence, for the power that comes from understanding the world and anticipating the actions of others, has driven the development of intricate systems of espionage, cryptography, and surveillance, shaping the course of human history and leaving a profound impact on the delicate balance between security and freedom.

This chapter embarks on a journey through the evolution of intelligence manufacturing and engineering, exploring the methods, motivations, and ethical implications of intelligence gathering, from the clandestine whispers of spies in the ancient world to the sophisticated algorithms and pervasive surveillance technologies of the modern digital age. We will delve into the shadowy world of intelligence operations, where secrets are currency, deception is a tool, and the line between protection and intrusion can become blurred.

The protection of intelligence has been a constant struggle, from the coded messages etched on clay tablets in ancient Mesopotamia to the encrypted communications that traverse the global internet. The methods of intelligence hacking have evolved alongside the methods of gathering, from the painstaking deciphering of ancient scripts to the sophisticated cyberattacks that can cripple nations in the modern era.

The motivations behind intelligence gathering have also shifted from the immediate needs of warfare and political maneuvering to the long-term goals of economic dominance and social control. As technology has advanced, so too has the potential for mass surveillance, raising profound ethical questions about the balance between security and individual liberties.

This chapter will explore these complex issues, examining the historical development of intelligence manufacturing and engineering, the evolving challenges of intelligence protection and hacking, and the ethical implications of increasingly sophisticated surveillance technologies. By understanding the past, we can better navigate the present and shape a future where intelligence serves the betterment of humanity, fostering cooperation, understanding, and progress rather than fueling conflict, oppression, and control.

INTELLIGENCE IN THE ANCIENT WORLD

In the ancient civilizations of Mesopotamia, Egypt, and Greece, intelligence gathering was a shadowy dance of espionage and statecraft, where information was a prized currency and secrecy was paramount. Rulers, ever vigilant against internal and external threats, relied on a network of spies and informants to weave a tapestry of knowledge about rival kingdoms, potential uprisings, and the whispers of discontent that could disrupt the delicate balance of power.

The value of intelligence lies not merely in the accumulation of facts but in its ability to provide a strategic advantage in the intricate game of politics and warfare. By deciphering the intentions of adversaries, anticipating their movements, and understanding the undercurrents of social unrest, rulers could maintain order, secure their borders, and safeguard their reign.

However, the very act of intelligence gathering created new vulnerabilities, exposing the secrets of states and individuals to the risk of interception, manipulation, and betrayal. The art of counterintelligence emerged as a vital companion to espionage, seeking to protect sensitive information from prying eyes and to mislead adversaries with misinformation and deception.

The challenges of intelligence protection and hacking were as old as civilization itself. Spies and informants were not only tasked with gathering secrets but also with safeguarding their own identities and communication channels. Codes and ciphers were developed to encrypt messages, while counterintelligence agents worked tirelessly to identify and neutralize enemy spies and disrupt their networks.

The stakes were high, as a compromised intelligence operation could lead to military defeat, political upheaval, or even the downfall of an entire civilization. The history of these ancient empires is littered with tales of espionage, betrayal, and the devastating consequences of intelligence failures.

The legacy of these early intelligence practices continues to shape the modern world, where the pursuit of information remains a driving force in politics, warfare, and commerce. Intelligence protection and hacking challenges have become even more complex in the digital age, as cyber espionage, data breaches, and misinformation campaigns threaten individuals, organizations, and even national security.

The development of writing systems, a monumental leap in human communication, enabled the recording and transmission of knowledge and gave rise to the intricate world of intelligence and counterintelligence. Cryptography, the art of secret writing, emerged as a vital tool for protecting sensitive information from prying eyes, allowing for the secure communication of military strategies, diplomatic negotiations, and trade secrets. However, the very existence of these coded messages sparked a parallel pursuit: the art of deciphering and intercepting them.

Even in these early societies, intelligence protection and hacking challenges were paramount. Rival factions, driven by the desire for power and advantage, employed skilled individuals to intercept messages, break codes, and manipulate information to their benefit. The art of deception and misinformation flourished, with cunning spies and double agents weaving intricate webs of deceit to mislead adversaries and protect their secrets.

This constant battle between those seeking to safeguard information and those determined to uncover it fueled a cycle of innovation and adaptation. New cryptographic techniques were developed to thwart the efforts of codebreakers while spies and counterspies engaged in a shadow dance of intrigue and deception. The stakes were high, as the intelligence compromise could lead to military defeats, diplomatic failures, and the loss of power and influence.

This early struggle for intelligence dominance foreshadows the complex challenges of cybersecurity in the modern digital age. The fundamental principles remain the same: the need to protect sensitive information, the persistent threat of malicious actors seeking to exploit vulnerabilities, and the constant evolution of techniques and technologies to safeguard against these threats.

THE RISE OF TECHNOLOGY AND INTELLIGENCE GATHERING

The invention of the printing press and the subsequent explosion of printed materials during the Renaissance irrevocably transformed the intelligence landscape. The ability to mass-produce written works democratized knowledge, empowering individuals and communities with unprecedented access to information. This newfound access fueled the Renaissance, sparked scientific revolutions, and ignited social and political movements. However, this democratization of information also presented new challenges for those seeking to control the flow of knowledge. Governments and institutions grappled with the implications of a more informed populace, leading to increased efforts to censor and restrict access to certain types of information.

The Industrial Revolution and the advent of electronic communication further accelerated the pace of intelligence gathering and dissemination. The telegraph, telephone, and radio enabled near-instantaneous communication across vast distances, revolutionizing warfare, diplomacy, and commerce. While these advancements connected the world in unprecedented ways, they also introduced new vulnerabilities. The ability to intercept and decipher electronic communications became a coveted tool for governments and corporations seeking strategic advantage. This sparked an escalating race between those seeking to secure communications and those determined to break those security measures.

The challenges of protecting sensitive information and ensuring the integrity of intelligence became increasingly complex. Cryptography evolved rapidly, with codes and ciphers becoming more sophisticated in an attempt to safeguard confidential communications. The rise of electronic espionage and signals intelligence agencies underscored the growing importance of cybersecurity in protecting national and economic interests.

INTELLIGENCE IN THE DIGITAL AGE

The rise of the internet and the proliferation of digital devices have ushered in an era of unprecedented interconnectedness, where information flows freely across borders and communities, transcending the limitations of physical distance and traditional communication channels. While offering immense opportunities for collaboration, innovation, and knowledge sharing, this hyper-connected digital landscape has amplified the challenges of intelligence protection and hacking, creating a complex and ever-evolving threat landscape.

Cyberattacks, once confined to the realm of technical experts and computer enthusiasts, have become increasingly sophisticated and pervasive, targeting individuals, organizations, and even nations with devastating consequences. Data breaches, exposing sensitive personal and financial information, can lead to identity theft, financial ruin, and the erosion of trust in digital systems. Misinformation campaigns, fueled by the rapid spread of false or misleading information through social media and online platforms, can manipulate public opinion, incite social unrest, and undermine democratic processes.

The very nature of intelligence protection and hacking has been transformed in this digital age. No longer confined to the physical realm of spies and intercepted documents, the battle for information now takes place in the ethereal realm of cyberspace, where malicious actors can exploit vulnerabilities in software, manipulate human psychology, and leverage the interconnectedness of the digital world to launch attacks with unprecedented speed and scale.

Protecting critical infrastructure, from power grids and financial systems to healthcare networks and transportation systems, has become a paramount concern. Cyberattacks targeting these vital systems can disrupt essential services, cause economic damage, and threaten national security. The Stuxnet worm, a sophisticated malware targeting Iranian nuclear facilities, is a stark reminder of the potential consequences of cyberattacks on critical infrastructure.

The rise of artificial intelligence and machine learning has further complicated the landscape of intelligence protection and hacking. While these technologies offer promising solutions for cybersecurity, they can also be exploited by malicious actors to develop more sophisticated and evasive attacks. The use of AI-powered tools for social engineering, phishing scams, and the creation of deepfakes highlights the evolving nature of cyber threats and the need for constant vigilance and adaptation.

In this complex and ever-changing digital environment, protecting intelligence and preserving trust in digital systems have become critical challenges for individuals, organizations, and governments alike. Developing robust cybersecurity measures, promoting digital literacy and awareness, and fostering a culture of ethical and responsible technology use are essential for navigating the complexities of the digital age and ensuring a secure and prosperous future for all.

Governments and corporations wield increasingly sophisticated surveillance technologies, casting a wide net to monitor online activities, amass vast troves of data, and exert influence over the flow of information. This pervasive surveillance, particularly acute in societies already burdened by strict Censorship and limited digital freedoms, raises profound ethical concerns about the erosion of privacy, the stifling of individual autonomy, and the potential for abuse of power.

The technologies that empower us to connect, communicate, and access information can also be employed to track our digital footprints, analyze our online behavior, and predict our future actions. This mass surveillance, often conducted under the guise of national security or public safety, can chill free speech, stifle dissent, and create an atmosphere of fear and self-censorship.

In societies where digital freedoms are already curtailed, the implications of such surveillance are even more dire. The ability to monitor online activities, intercept communications, and track individuals' movements can be used to suppress opposition, silence critical voices, and maintain authoritarian control. The lack of transparency and accountability surrounding these surveillance practices further exacerbates the potential for abuse of power, leaving individuals vulnerable to arbitrary detention, persecution, and the violation of their fundamental rights.

The erosion of privacy in the digital age extends beyond government surveillance. Corporations, driven by the pursuit of profit, collect vast amounts of personal data, often without the informed consent of individuals. This data creates detailed profiles, predicts consumer behavior, and targets individuals with personalized advertising. The lack of control over one's data and the potential for misuse raises serious concerns about autonomy, self-determination, and the very nature of identity in the digital age.

As technology advances and the lines between the physical and digital worlds blur, the ethical implications of surveillance and data collection become increasingly complex and urgent. Safeguarding privacy, preserving individual autonomy, and ensuring accountability in using surveillance technologies are crucial challenges that demand our attention. The future of a free and open society hinges on our ability to navigate these complexities and strike a balance between the benefits of technology and the fundamental rights of individuals.

THE FUTURE OF INTELLIGENCE

In the face of increasing technological control and Censorship, intelligence's future may reside in the hands of resilient communities and individuals who leverage technology to defend their digital freedoms and champion the open sharing of knowledge. The development of privacy-enhancing technologies, decentralized platforms, and secure communication channels could empower individuals to circumvent Censorship, resist the encroachment of surveillance, and maintain control over their information. This struggle for digital autonomy represents a crucial battleground in the ongoing tug-of-war between control and freedom, between the forces of surveillance and the spirit of open inquiry.

The essence of intelligence – the ability to gather, analyze, and disseminate information – is threatened in an environment where governments and corporations increasingly seek to control the flow of knowledge and manipulate narratives. Cybersecurity, therefore, becomes a critical component in safeguarding not only individual privacy but also the collective intelligence of a society. Encryption technologies, anonymization tools, and secure communication protocols act as digital shields, protecting sensitive information from prying eyes and allowing for the free exchange of ideas even in the face of Censorship and surveillance.

Decentralized platforms, operating independently of centralized control, offer alternative spaces for expression, collaboration, and the dissemination of information. These platforms challenge the dominance of monolithic social media giants and allow diverse voices and perspectives to flourish.

The future of intelligence hinges on the ability of individuals and communities to harness the empowering potential of technology while remaining vigilant in safeguarding their digital freedoms. The societies that thrive in this dynamic environment will foster a culture of critical engagement with information, encouraging citizens to question, analyze, and evaluate the information they encounter. They will be societies that recognize the importance of digital literacy, equipping individuals with the skills and knowledge to navigate the complex digital landscape and discern truth from falsehood.

Ultimately, the future of intelligence will be shaped by the ongoing struggle between those who seek to control information and those who champion the free flow of knowledge. The societies that embrace the empowering potential of technology while remaining steadfast in their commitment to individual liberties and open inquiry will be the ones that thrive in the digital age and beyond.

TUG-OF-WAR: NAVIGATING THE COMPLEX LANDSCAPE OF DATA SECURITY

In today's hyper-connected world, data has become a currency more valuable than gold, fueling innovation, driving economic growth, and shaping the trajectory of societal progress. Nevertheless, this data-driven revolution has also given rise to a complex and often contentious landscape of intelligence security, where corporations, governments, and individuals are locked in a constant tug-of-war over the control, access, and protection of sensitive information.

Corporations, driven by the insatiable thirst for data to refine their products, target their marketing strategies and gain a competitive edge, relentlessly gather information, often pushing the boundaries of personal privacy. This relentless pursuit of data has led to the development of sophisticated tracking technologies, the aggregation of massive datasets, and the rise of predictive analytics, raising concerns about the erosion of privacy and the potential for misuse of personal information.

Governments, recognizing the strategic value of data, also play a prominent role in this landscape. They grapple with balancing national security concerns with protecting individual liberties, navigating the complex terrain of surveillance, data interception, and data flow regulation across borders. The tension between the need to protect sensitive information from foreign adversaries and the desire to foster innovation and economic growth in the digital age creates a delicate balancing act for policymakers.

Individuals caught in the crossfire of this struggle find themselves increasingly concerned about the security and privacy of their data. The rise of data breaches, identity theft, and online scams has heightened awareness of the vulnerabilities inherent in the digital world. The desire to protect personal information while enjoying the benefits of a connected world creates constant tension for individuals navigating the digital landscape.

This chapter delves into the intricate dynamics of this struggle, examining the motivations, strategies, and ethical implications of data collection, storage, and utilization in the digital age. It explores the complex interplay between corporations, governments, and individuals, highlighting the challenges of balancing innovation, security, and privacy protection in a world where data has become the new gold.

THE DATA GOLD RUSH: CORPORATIONS AND THE QUEST FOR INFORMATION

Corporations have become voracious data consumers in the digital age, recognizing their immense value in understanding consumer behavior, tailoring marketing strategies, and developing innovative products and services. The quest for data has led corporations to employ increasingly sophisticated techniques to gather information, often blurring the lines between legitimate data collection and intrusive surveillance. This relentless pursuit of data has raised concerns about the erosion of privacy and the potential to misuse sensitive information.

From tracking online browsing habits and social media interactions to analyzing purchasing patterns and location data, corporations leave no digital stone unturned in their pursuit of valuable insights. The rise of the Internet of Things (IoT), with its interconnected web of smart devices, has expanded the scope of data collection, providing corporations with a treasure trove of information about our homes, habits, and health. When aggregated and analyzed, this data can reveal intimate details about our lives, preferences, and vulnerabilities.

The aggregation of vast troves of personal data in the hands of corporations creates significant cybersecurity risks. Data breaches, whether caused by malicious attacks or unintentional leaks, can expose sensitive information to unauthorized access, leading to identity theft, financial fraud, and reputational damage. The potential for misuse of this data is vast, ranging from targeted advertising and manipulation to discriminatory practices based on individuals' digital profiles.

Moreover, the lack of transparency and control over how corporations collect, store, and utilize personal data raises ethical concerns. Individuals often have little say in how their data is used, and the complex algorithms employed to analyze this data can perpetuate biases and reinforce existing inequalities.

The challenge lies in balancing the benefits of data-driven innovation and protecting individual privacy and security. This requires a multifaceted approach, including robust data protection regulations, increased corporate transparency, and the development of privacy-enhancing technologies that empower individuals to control their data.

Furthermore, fostering a culture of cybersecurity awareness is essential. Individuals need to be educated about the risks associated with data collection and the importance of safeguarding their personal information. Corporations must prioritize data security, implementing robust cybersecurity measures to protect against breaches and misuse.

The responsible use of data in the digital age is a complex and evolving challenge. By promoting transparency, accountability, and individual empowerment, we can harness the benefits of data-driven innovation while safeguarding privacy, security, and the ethical use of information.

THE REGULATORY TIGHTROPE: BALANCING SECURITY AND THE DATA MARKET

In response to escalating anxieties surrounding data privacy and security, governments and regulatory bodies have stepped into the arena, attempting to strike a delicate balance between safeguarding sensitive information and fostering the burgeoning data market. Regulations such as the General Data Protection Regulation (GDPR) in Europe and the California Consumer Privacy Act (CCPA) in the United States aim to empower individuals with greater control over their personal data while providing a framework for corporate responsible data use. These regulations, though well-intentioned, often struggle to keep pace with the rapid evolution of technology and the ever-shifting tactics of corporations eager to exploit data for profit.

The complexity of data privacy laws, coupled with the internet's borderless nature, creates enforcement challenges and a patchwork of regulations that can be difficult for corporations and individuals to navigate. This complexity can inadvertently create loopholes and inconsistencies, leaving individuals vulnerable to data exploitation and hindering the development of a unified global approach to data protection. Moreover, the rapid pace of technological advancement often outstrips the ability of regulatory frameworks to adapt, leaving emerging technologies and data practices unregulated and open to abuse.

The struggle to protect data privacy and security in the digital age is a dynamic and ongoing challenge. It requires a multifaceted approach encompassing robust regulations, technological innovation, cybersecurity awareness, and international cooperation. Empowering individuals with the knowledge and tools to protect their data, fostering ethical data practices among corporations, and promoting a global dialogue on data governance are crucial steps toward creating a more secure and equitable digital world.

THE USER'S DILEMMA: NAVIGATING A COMPLEX LANDSCAPE

In this intelligence tug-of-war, individuals often find themselves caught in a web of vulnerability and disempowerment. The sheer volume of data collected about them, coupled with the opaque nature of data collection practices and the labyrinthine complexity of privacy regulations, can make it exceedingly challenging to comprehend and control how their personal information is used. This lack of transparency and control can erode trust, foster a sense of powerlessness, and create fertile ground for exploitation.

The constant barrage of consent requests, privacy policies, and data breach notifications can lead to a phenomenon known as "consent fatigue." Individuals become desensitized to protecting their data, overwhelmed by the sheer volume of requests and the often incomprehensible legalese of privacy policies. This fatigue can have serious consequences, making individuals more susceptible to manipulative marketing tactics, social engineering scams, and other online exploitation that prey on diminished vigilance.

Furthermore, the increasing reliance on artificial intelligence and machine learning in data analysis raises legitimate concerns about algorithmic bias and the potential for discriminatory practices. The lack of transparency in how these algorithms are developed and deployed can perpetuate existing inequalities and create new forms of digital discrimination, further marginalizing vulnerable communities and eroding trust in digital systems.

This erosion of trust has profound implications for the future of societal intelligence. When individuals feel disempowered and unable to control their data, they may become disengaged from online platforms and critical of technological advancements. This disengagement can hinder the free flow of information, stifle innovation, and create barriers to collaboration, ultimately undermining the collective intelligence of a society.

The landscape of intelligence security in today's society is complex and ever-evolving. The relentless pursuit of data by corporations, driven by the desire to monetize user information and gain a competitive edge, often clashes with the efforts of regulations designed to protect individuals' privacy and digital rights. This creates a dynamic and often contentious arena where the boundaries between innovation and exploitation, legitimate data use, and privacy violations are constantly being negotiated and redefined.

Cybersecurity threats loom large in this landscape as malicious actors seek to exploit system vulnerabilities and human behavior to gain unauthorized access to sensitive information. Data breaches, ransomware attacks, and misinformation campaigns pose significant risks to individuals, organizations, and national security.

The challenges faced by users in navigating this complex environment are multifaceted. The sheer volume of data collected, the often opaque practices of data brokers, and the increasing sophistication of cyberattacks can leave individuals feeling overwhelmed and vulnerable. The erosion of privacy, the potential for manipulation, and the misuse of personal information for targeted advertising or even discriminatory practices raise profound ethical and societal concerns.

As technology advances and data's importance grows, the need for greater transparency, accountability, and user empowerment will become increasingly critical. Corporations must be held accountable for their data collection and use practices, prioritizing user privacy and data security. Regulations must evolve to keep pace with rapid technological advancements, providing clear guidelines and safeguards to protect individuals' digital rights.

Users, in turn, must be empowered with the knowledge and tools to navigate the digital landscape safely and responsibly. Digital literacy, critical thinking skills, and an understanding of cybersecurity threats are essential for individuals to protect their privacy, make informed choices about their online activities, and participate fully in the digital age.

The societies that thrive in this data-driven world will strike a delicate balance between innovation and privacy, ensuring that all benefits of technology are shared while safeguarding the fundamental rights and freedoms of individuals in the digital age. This requires a collective effort involving governments, corporations, and individuals to create a digital ecosystem that fosters innovation, protects privacy, and empowers users to navigate the complex landscape of intelligence security with confidence and resilience.

CONTROLS AND CENSORSHIP, ADVERSARIAL

Throughout history, the siren song of power has tempted governments and ruling authorities to employ various forms of societal control, seeking to maintain order, enforce conformity, and quell the flames of dissent. These controls, ranging from the blatant iron fist of Censorship and propaganda to the subtle manipulation of social engineering, weave a tapestry of influence that has profound and often insidious impacts on individuals, communities, and the very fabric of society. This chapter delves into the intricate web of societal controls, dissecting their underlying mechanisms and analyzing their potential to sow confusion, foster distrust, and cultivate fertile ground for adversarial exploitation, particularly in cybersecurity and intelligence.

The control methods are as varied as the societies that employ them. With its heavy hand, Censorship seeks to silence dissenting voices and restrict the free flow of information, creating an illusion of consensus while suppressing alternative perspectives. Propaganda, the art of persuasion through manipulation, twists narratives, distorts facts, and preys on emotions to mold public opinion and manufacture consent. Social engineering, a more subtle form of control, exploits psychological vulnerabilities and social dynamics to influence behavior, often to gain access to sensitive information or compromise security systems.

The consequences of these controls extend far beyond the immediate suppression of dissent or the manipulation of information. They erode trust, sow division, and create an atmosphere of fear and suspicion where individuals hesitate to express their accurate opinions or challenge the prevailing narrative. This climate of intellectual repression not only stifles innovation and progress but also leaves societies vulnerable to adversarial attacks, both from within and without.

In cybersecurity, societal controls can create a fertile ground for malicious actors to exploit. Censorship and the suppression of information can hinder the development of a robust cybersecurity culture, leaving individuals and organizations ill-equipped to recognize and respond to cyber threats. The lack of open dialogue about vulnerabilities and attack vectors can create a false sense of security, while the manipulation of information can be used to spread disinformation, sow discord, and undermine trust in digital systems and institutions.

Moreover, the erosion of trust and the fostering of suspicion within a society can make detecting and responding to intelligence compromises more difficult. When individuals are hesitant to share information or report suspicious activity for fear of reprisal or surveillance, it creates blind spots that adversaries can exploit. The chilling effect of Censorship can also discourage whistleblowers and investigative journalists from exposing wrongdoing, further hindering the ability to identify and address threats.

In conclusion, the various forms of societal control employed throughout history have profound implications for cybersecurity and intelligence. By understanding the mechanisms and consequences of these controls, we can develop strategies to resist manipulation, protect our digital freedoms, and foster a culture of critical engagement with information. In an increasingly interconnected world, where the boundaries between the physical and digital realms are becoming increasingly blurred, the ability to discern truth from falsehood, resist adversarial exploitation, and maintain a resilient and informed society is paramount.

INFORMATION CONTROL: SUPPRESSING
THE FREE FLOW OF KNOWLEDGE

Information manipulation is one of the most potent weapons in the arsenal of societal control. Governments and powerful institutions throughout history and across the globe have recognized the immense power of controlling the narrative, shaping public opinion, and limiting access to information that might challenge their authority or disrupt the established order. Censorship, in its myriad forms, acts as a muzzle, silencing dissenting voices and suppressing the free flow of ideas that could spark change or challenge the status quo. Propaganda, with its seductive blend of half-truths, distortions, and outright falsehoods, molds public perception, manufacturing consent for policies and actions that might otherwise be met with resistance. The suppression of dissenting voices through intimidation, persecution, or the subtle chilling effect of self-censorship creates an illusion of consensus, silencing those who dare to question or challenge the prevailing narrative.

In the digital age, the manipulation of information has taken on new dimensions with the rise of sophisticated technologies that enable the surveillance, tracking, and manipulation of online discourse. Algorithms curate our news feeds, shaping our perceptions of reality and reinforcing existing biases. Social media platforms, while offering powerful tools for connection and expression, can also be weaponized to spread disinformation, sow discord, and manipulate public opinion.

The consequences of information control are far-reaching, eroding trust in institutions, undermining social cohesion, and creating fertile ground for adversarial exploitation. When citizens are denied access to accurate and unbiased information, they become vulnerable to manipulation, susceptible to propaganda, and less equipped to engage in critical thinking and informed decision-making. This vulnerability can be readily exploited by malicious actors seeking to sow discord, spread misinformation, and undermine democratic processes.

Therefore, the struggle against information control is not merely a battle for freedom of expression but a fight to preserve the foundation of an informed and resilient society. It is a fight to protect the individual's right to access information, to question authority, and to engage in open and critical dialogue. In an increasingly interconnected world, where the flow of information transcends borders, and the digital landscape shapes our perceptions of reality, the ability to discern truth from falsehood, resist manipulation, and engage in informed decision-making is paramount for the preservation of individual liberties, the promotion of societal well-being, and the advancement of human progress.

The Soviet Union's iron grip on information flow serves as a stark example of how Censorship and the suppression of dissent can cripple a society's intellectual growth and leave it vulnerable to manipulation. By censoring publications, controlling the media, and persecuting those who dared to challenge the official ideology, the Soviet regime created an atmosphere of fear and mistrust that permeated every aspect of life. Individuals became hesitant to express their accurate opinions, even in private, and the vital exchange of ideas that fuels innovation and progress ground to a halt.

This intellectual stagnation had profound consequences, limiting scientific and cultural advancement and leaving the population acutely vulnerable to propaganda and misinformation. With limited access to diverse perspectives and independent sources of information, citizens became reliant on state-controlled media, which skillfully manipulated narratives to maintain control and suppress dissent.

The lack of critical thinking skills, cultivated in an environment where questioning the official line was met with persecution, further exacerbated this vulnerability. Citizens became less equipped to discern truth from falsehood, leaving them susceptible to the manipulative tactics of the regime. This hindered the development of a robust and informed public discourse and created fertile ground for the spread of disinformation and propaganda, further solidifying the regime's control.

The Soviet Union's experience serves as a cautionary tale of the perils of Censorship and the importance of safeguarding the free flow of information. In an increasingly interconnected world, where cyber threats and misinformation campaigns can readily transcend borders, accessing diverse perspectives, engaging in critical thinking, and discerning truth from falsehood are paramount for the security and well-being of individuals and societies alike.

SOCIAL CONTROL: ENGINEERING CONFORMITY AND OBEDIENCE

Social control mechanisms, woven into the fabric of societies throughout history, influence individual behavior, shaping actions, beliefs, and even thoughts to enforce conformity and uphold established norms. These mechanisms operate through a complex interplay of overt and subtle tactics, ranging from the pervasive pressure of social expectations and the chilling effect of ostracization to the intrusive gaze of surveillance technologies and the heavy hand of legal sanctions against dissent.

In the digital age, social control mechanisms have extended their reach into the virtual realm, where the ubiquitous presence of technology and the interconnectedness of online spaces provide new avenues for monitoring, manipulating, and shaping individual behavior. Social media platforms, with their intricate algorithms and vast troves of user data, can be leveraged to amplify social pressure, promote conformity, and subtly nudge individuals toward desired behaviors.

The rise of surveillance technologies, from facial recognition software to sophisticated data mining techniques, further enhances the capacity for social control. The constant monitoring of online activities, the tracking of digital footprints, and the analysis of personal data create an environment where individuals may feel pressured to conform, self-censor, and suppress dissenting opinions for fear of social or legal repercussions.

The criminalization of dissent, through laws restricting freedom of expression or punishing those challenging the status quo, represents a more overt form of social control. Such measures not only silence dissenting voices but also create a chilling effect, discouraging individuals from expressing their true beliefs or engaging in critical thinking.

The consequences of these social control mechanisms extend far beyond the suppression of individual freedoms. They can erode trust, foster paranoia, and create fear and self-censorship, hindering creativity, innovation, and social progress. In cybersecurity, this erosion of trust can have particularly detrimental effects, making individuals more susceptible to social engineering attacks, phishing scams, and other forms of online manipulation.

When individuals feel pressured to conform and suppress their true beliefs, they may be less likely to question suspicious emails, challenge authority figures, or report potential security breaches. This can create vulnerabilities in personal and organizational cybersecurity, making it easier for malicious actors to exploit the trust deficit and access sensitive information.

Moreover, the pervasive surveillance and monitoring of online activities can create a chilling effect on cybersecurity research and innovation. When individuals fear that their work may be scrutinized or censored, they may be less inclined to explore controversial topics, challenge established norms, or develop innovative solutions that could disrupt the status quo.

In conclusion, social control mechanisms, whether operating through social pressure, surveillance technologies, or the criminalization of dissent, have profound implications for cybersecurity and the resilience of societies in the digital age. By fostering a culture of open dialogue, critical thinking, and digital literacy, we can empower individuals to resist manipulation, protect their digital freedoms, and contribute to a more secure and resilient society.

China's social credit system epitomizes the chilling potential for technology to be wielded as a tool of social control, blurring the lines between technological advancement and Orwellian dystopia. With its intricate web of surveillance and social engineering, this sophisticated system tracks citizens' behavior with an unblinking eye, assigning social credit scores that dictate access to essential services and societal privileges. By rewarding adherence to social norms and punishing deviations, the system insidiously aims to engineer a society of conformity and obedience, where individuals are constantly monitored and evaluated, their every move scrutinized and quantified.

The consequences of such a system extend far beyond the erosion of privacy and individual autonomy. The social credit system has the potential to create a deeply stratified society, where those deemed "trustworthy" by the authorities enjoy preferential treatment, while those who fall short face restrictions, sanctions, and social ostracism. This stifles dissent and critical thinking and fosters an environment of fear and self-censorship, where individuals are hesitant to express their accurate opinions or challenge the prevailing orthodoxy.

Furthermore, the social credit system raises profound concerns about potential abuse and manipulation. The opaque nature of the scoring system, coupled with the lack of transparency and accountability, leaves individuals vulnerable to arbitrary judgments and the potential for political persecution. The system could be readily exploited to silence critics, suppress dissent, and maintain a tight grip on power.

In the context of cybersecurity and intelligence compromise, the social credit system represents a formidable threat. The vast amounts of data collected on citizens' behavior, online activities, and social interactions create a treasure trove of information that malicious actors could exploit. Data breaches, leaks, or unauthorized access to this sensitive information could have devastating consequences, enabling identity theft, targeted harassment, or even the manipulation of social credit scores for malicious purposes.

Moreover, the social credit system's reliance on technology creates new cyberattack vulnerabilities. Disrupting or manipulating the system's infrastructure could sow chaos, erode trust in the government, and create opportunities for adversaries to exploit the resulting instability.

The Chinese social credit system serves as a stark warning about the potential for technology to be used for social control and the erosion of fundamental freedoms. It highlights the importance of vigilance, ethical considerations in technology development, and the need to safeguard individual liberties and digital rights in the face of increasingly sophisticated surveillance and social engineering technologies.

PSYCHOLOGICAL CONTROL: MANIPULATING MINDS AND EMOTIONS

Psychological control tactics are not merely tools of persuasion; they are instruments of manipulation designed to influence individuals' thoughts, emotions, and behaviors through subtle and often insidious means. These tactics, employed by governments, institutions, and even individuals, seek to instill fear, loyalty, and unquestioning obedience, ultimately undermining critical thinking and independent judgment.

Propaganda, a cornerstone of psychological control, distorts information, manipulates narratives, and appeals to emotions to shape public opinion and manufacture consent. Fear-mongering, the deliberate dissemination of anxiety and apprehension, creates a climate of insecurity, making individuals more susceptible to manipulation and control. The creation of "enemies," whether real or imagined, fosters a sense of threat and division, justifying the suppression of dissent and the consolidation of power.

The insidious nature of psychological control lies in its ability to bypass conscious awareness, subtly shaping beliefs and behaviors through repeated exposure to manipulative messages and emotional appeals. This can lead to self-censorship, conformity, and a reluctance to question authority, ultimately eroding individual autonomy and critical thinking.

The consequences of psychological control can be devastating, leading to the suppression of dissent, the erosion of trust, and the acceptance of authoritarianism. In extreme cases, it can pave the way for genocide and other atrocities as individuals become desensitized to violence and dehumanization.

Recognizing and resisting psychological control tactics is essential for preserving individual freedoms, fostering critical thinking, and promoting a just and equitable society. By cultivating media literacy, encouraging open dialogue, and challenging manipulative narratives, we can build resilience against psychological manipulation and safeguard our minds and communities' integrity.

The Nazi regime, under the sinister orchestration of Joseph Goebbels, wielded propaganda as a weapon of mass manipulation, expertly exploiting the vulnerabilities of human psychology to mold public opinion and pave the way for its horrific agenda. Through a carefully crafted symphony of emotionally charged rhetoric, the demonization of minority groups, and the relentless glorification of Nazi ideology, the regime systematically eroded critical thinking, stoked fear and hatred, and ultimately engineered a society complicit in its atrocities.

Goebbels, a master of propaganda, understood the power of appealing to primal emotions. He skillfully crafted messages that tapped into the anxieties and frustrations of the German people, offering scapegoats in the form of minority groups and promising a utopian future under Nazi rule. Through mass rallies, controlled media, and even the education system, the relentless repetition of these messages created an echo chamber where dissenting voices were silenced and critical thinking was replaced by blind obedience.

The demonization of minority groups, particularly Jews, Roma, and homosexuals, served to create a common enemy, a scapegoat upon which to project societal anxieties and frustrations. This dehumanization of entire groups not only fueled hatred and prejudice but also paved the way for the regime's policies of persecution, segregation, and, ultimately, genocide.

The glorification of Nazi ideology, with its emphasis on racial purity, national unity, and the cult of the Führer, created a seductive narrative that promised a return to greatness and a restoration of German pride. This narrative, skillfully interwoven with propaganda and reinforced through the suppression of alternative viewpoints, created a powerful illusion of consensus, masking the regime's true intentions and silencing any opposition.

The consequences of this masterful manipulation of public opinion were devastating. Millions of individuals, swayed by the relentless barrage of propaganda, became complicit in the regime's atrocities, either through active participation or passive acquiescence. The horrors of the Holocaust, the systematic extermination of millions of innocent people, stand as a chilling testament to the destructive power of propaganda and the vulnerability of human psychology to manipulation and control.

THE CONSEQUENCES OF CONTROL: SOCIAL CONFUSION AND ADVERSARIAL EXPLOITATION

The cumulative impact of these societal controls, like layers of sediment slowly burying a vibrant ecosystem, creates a suffocating climate of confusion, distrust, and vulnerability. When individuals are denied access to accurate and unbiased information, subjected to the relentless manipulation of narratives, and constantly monitored by the watchful eyes of surveillance, their capacity for critical thinking erodes, leaving them susceptible to adversarial exploitation in both the physical and digital realms.

This vulnerability manifests in myriad ways. Citizens, starved of factual information and bombarded with propaganda, struggle to discern truth from falsehood, making them easy prey for misinformation campaigns and online manipulation. The chilling effect of surveillance discourages dissent and critical discourse, creating an intellectual vacuum where alternative perspectives are silenced, and the seeds of doubt are sown.

In the digital sphere, this vulnerability translates into a heightened risk of cyber-attacks. Individuals, conditioned to accept information passively and discouraged from questioning authority, become prime targets for phishing scams, social engineering ploys, and other forms of cyber manipulation. The lack of open dialogue about cybersecurity threats and the suppression of information about vulnerabilities further exacerbates this risk.

Moreover, the erosion of trust accompanying pervasive societal control creates fertile ground for adversarial exploitation. When individuals lose faith in institutions, media sources, and even their communities, they become more susceptible to the divisive tactics of malicious actors seeking to sow discord and undermine social cohesion.

In essence, the cumulative impact of societal controls creates a society ripe for exploitation, where individuals are vulnerable to manipulation and control and ill-equipped to defend themselves against the ever-evolving threats of the digital age. The suppression of information, the stifling of critical thinking, and the erosion of trust weaken the very fabric of society, leaving it exposed to the insidious forces that seek to undermine its integrity and exploit its vulnerabilities.

Malicious actors thrive in an environment stifled by societal controls, where the free flow of information is choked and critical thinking is suppressed. Like opportunistic pathogens exploiting a weakened immune system, foreign governments, extremist groups, and cybercriminals readily capitalize on the vulnerabilities created by these controls. The spread of misinformation becomes rampant, weaving its tendrils through the fabric of society, distorting reality, and sowing seeds of discord. Public opinion, malleable in the absence of diverse perspectives and open debate, is easily swayed and manipulated, becoming a tool for those seeking to advance their agendas.

The disruption of critical infrastructure, the backbone of modern society, becomes a chillingly achievable goal. Power grids, financial systems, and communication networks, weakened by the lack of transparency and open collaboration, are left exposed to cyberattacks that can cripple essential services, sow chaos, and erode public trust.

This vulnerability stems from a fundamental erosion of societal intelligence. When individuals lack the tools and knowledge to discern truth from falsehood, to critically evaluate information, and to resist manipulation, they become pawns in a larger game. The foundations of a resilient society – informed decision-making, critical thinking, and a shared understanding of reality – crumble under the weight of Censorship and control.

The consequences are far-reaching, impacting individual liberties, national security, and global stability. In a world increasingly reliant on interconnected digital systems, the compromise of intelligence in one society can have cascading effects, rippling across borders and destabilizing entire regions.

Therefore, safeguarding against these threats requires a multi-pronged approach. Fostering a digital literacy and critical thinking culture is paramount, empowering individuals to navigate the complex information landscape and resist manipulation. Promoting transparency and open dialogue about cyber threats is crucial, enabling collaboration and collective defense against malicious actors. Furthermore, perhaps most importantly, defending digital freedoms and resisting the urge to control and censor information is essential for building a resilient and informed society capable of withstanding the challenges of the digital age.

The landscape of societal control is a complex and ever-evolving tapestry woven with threads of manipulation, surveillance, and the suppression of dissent. From the overt Censorship of information to the subtle manipulation of minds and emotions, these controls profoundly influence individuals, communities, and the very fabric of society. By understanding the intricate mechanisms and far-reaching consequences of societal controls, we can begin to develop strategies to resist manipulation, safeguard our digital freedoms, and foster a culture of critical engagement with information.

In an increasingly interconnected world, where the flow of information transcends geographical boundaries, and digital technologies permeate every aspect of our lives, the ability to discern truth from falsehood and resist adversarial exploitation becomes paramount. The manipulation of information, the spread of disinformation, and the erosion of trust pose significant threats to the stability of societies and the well-being of individuals.

Cybersecurity, in this context, takes on a new dimension, extending beyond the protection of digital infrastructure to encompass safeguarding our cognitive freedoms and preserving societal intelligence. Encryption technologies, anonymization tools, and secure communication platforms become essential tools in the fight against Censorship and surveillance, empowering individuals and communities to resist control and maintain access to unfiltered information.

Cultivating critical thinking skills and media literacy becomes crucial in navigating the complex digital landscape. Evaluating information sources, identifying bias, and discerning fact from fiction are essential for informed decision-making and resisting manipulation.

Furthermore, fostering a culture of open dialogue and the free exchange of ideas is vital in countering the chilling effects of Censorship and control. By encouraging diverse perspectives, promoting critical engagement with information, and challenging dominant narratives, we can create a more resilient and informed society better equipped to resist adversarial exploitation and safeguard individual liberties.

In essence, the struggle against societal control is not merely a battle for access to information; it is a fight for preserving human autonomy, promoting critical thinking, and advancing human progress. By understanding the mechanisms and consequences of societal controls, we can empower individuals and communities to resist manipulation, protect their digital freedoms, and foster a society where knowledge is shared freely and the pursuit of truth remains unfettered.

The evolution of societal intercommunal intelligence has been a long and complex journey, shaped by the interplay of communication technologies, social structures, and the ever-present challenges of security and trust. From the slow and deliberate pace of ancient messengers to the instantaneous flow of information in the hyperconnected digital age, how communities gather and share knowledge has undergone a profound transformation, with significant implications for cybersecurity and the potential for intelligence compromise.

In the early days of civilization, when communication was constrained by the limitations of human messengers and handwritten correspondence, the pace of intercommunal intelligence was slow and deliberate. Knowledge was carried by envoys, merchants, and spies along trade routes and diplomatic channels. The challenges of intelligence gathering were immense, relying on the cunning of spies, deciphering coded messages, and interpreting often unreliable reports. However, the very slowness of communication also provided a degree of security. Information was complex to intercept and even more challenging to alter or manipulate on a large scale. Community trust was often built on long-standing relationships and personal interactions, making it harder for malicious actors to sow discord or spread misinformation.

The invention of the printing press marked a turning point, accelerating the dissemination of knowledge and fostering a new era of intellectual exchange. Books, pamphlets, and newspapers became vehicles for the spread of ideas, challenging traditional hierarchies and fueling social and political movements. However, this increased speed and volume of information also brought new vulnerabilities. The ability to mass-produce printed materials made it easier to spread propaganda and misinformation, while the centralized nature of printing presses made them susceptible to Censorship and control.

The advent of the telegraph, telephone, and radio ushered in a new era of instant communication, further accelerating the pace of intercommunal intelligence and connecting the world in unprecedented ways. However, with this increased connectivity came new challenges for cybersecurity. The ability to intercept and decipher electronic communications created new opportunities for espionage and sabotage, while the reliance on centralized infrastructure made communication networks vulnerable to disruption and attack.

Today, in the hyper-connected digital age, the flow of information has become a torrent, overwhelming traditional boundaries and creating a global village where communities are intertwined in ways never imagined. The internet, social media, and mobile devices have revolutionized the way we communicate, learn, and interact with each other, creating both opportunities and challenges for societal intelligence.

The interconnectedness of the digital age has blurred the lines between communities, making it harder to identify the source and assess the reliability of information. The potential for intelligence compromise is amplified by the speed and scale of information dissemination, making it easier for malicious actors to spread disinformation, manipulate public opinion, and disrupt social cohesion.

As we navigate this complex and ever-evolving digital landscape, the preservation of trust and the ability to discern truth from falsehood become paramount. Cybersecurity measures, media literacy education, and critical thinking skills are essential in safeguarding information integrity and ensuring that intelligence flows between communities remain secure and trustworthy.

In the nascent stages of human civilization, intelligence acquisition relied heavily on the clandestine activities of spies, deciphering encrypted messages,deciphering encrypted messages, and careful interpretation of reports often shrouded in ambiguity and misinformation. The invention of the printing press, while ushering in an era of unprecedented knowledge dissemination, also introduced new challenges to the security and control of information. Suddenly, ideas could spread rapidly beyond the confines of courts and confidential whispers, creating opportunities and threats for those seeking to wield the power of knowledge.

This nascent era of information warfare saw the rise of new forms of intelligence hacking. Coded messages, once the domain of diplomats and spies, became increasingly sophisticated, requiring more ingenious decryption methods. The control of information itself became a strategic objective, as those in power sought to manipulate public opinion, suppress dissent, and maintain their grip on authority. While a powerful tool for enlightenment, the printing press also became a weapon for those seeking to control the narrative and shape the course of history.

The advent of electronic communication, from the telegraph to the telephone and radio, further accelerated the pace of intercommunal intelligence, connecting the world in unprecedented ways and forever altering the landscape of information exchange. However, this newfound interconnectedness also amplified the challenges of securing communication channels and protecting sensitive information from interception and manipulation. The ethereal nature of electronic signals, traversing vast distances through the airwaves, made them susceptible to eavesdropping and interception by those with the technological prowess to tap into these invisible data streams.

The rise of cryptography as a science became intertwined with the evolution of communication technologies as individuals and nations sought to protect their secrets from prying eyes. Codes and ciphers grew increasingly complex, driving an arms race between those seeking to encrypt information and those determined to break those codes. The stakes were high, as the compromise of sensitive intelligence could have devastating consequences for individuals, organizations, and even nations.

Today, in the hyper-connected digital age, the flow of information has become a torrential flood, surging across traditional boundaries and forging a global village where communities intertwine in ways never imagined. The ease with which we share information and the lightning-fast dissemination of knowledge have undoubtedly fostered collaboration, innovation, and cultural exchange on a global scale. However, this interconnectedness comes at a cost, as the challenges of misinformation, cyberattacks, and the erosion of privacy have become increasingly acute.

The extreme difficulty of understanding the compromise of intelligence in this interconnected world poses a significant challenge to individuals, organizations, and nations alike. The technologies that connect us also expose us to new vulnerabilities, making it increasingly difficult to discern information's origins, intentions, and trustworthiness. Cybersecurity threats loom large, with malicious actors exploiting vulnerabilities to steal data, disrupt critical infrastructure, and spread misinformation.

The ease with which information can be manipulated and disseminated in the digital age makes it a powerful weapon for those seeking to sow discord, manipulate public opinion, or undermine trust in institutions. Deepfakes, sophisticated AI-generated forgeries, can blur the lines between reality and fabrication, making it increasingly difficult to distinguish authentic content from malicious propaganda.

The erosion of privacy further exacerbates the challenges of intelligence protection. As our digital footprints expand, vast amounts of personal data are collected, analyzed, and potentially exploited by governments, corporations, and malicious actors. This erosion of privacy undermines individual autonomy and creates new vulnerabilities for social engineering, targeted attacks, and behavior manipulation.

In this complex and ever-evolving digital landscape, critically evaluating information, discerning truth from falsehood, and protecting one's digital identity becomes paramount. Cybersecurity awareness, media literacy, and critical thinking skills are essential tools for navigating the challenges of the hyper-connected world and safeguarding the integrity of intelligence.

In boxed-up communities, where physical and digital boundaries restrict the free flow of information, the struggle for open knowledge and the protection of digital freedoms will be paramount. The societies that thrive in this environment will recognize the vital importance of cybersecurity, empowering individuals and communities with the tools and knowledge to navigate the digital landscape safely and responsibly. They will champion the development of privacy-enhancing technologies, promote digital literacy, and foster a culture of critical inquiry, ensuring that the pursuit of knowledge remains a cornerstone of human progress.

The journey of societal intercommunal intelligence is a testament to the human desire for connection, understanding, and progress. From the earliest exchanges of knowledge through cave paintings and oral traditions to the instantaneous communication of the digital age, humanity has relentlessly sought to bridge divides, share insights, and build upon the collective wisdom of our species. As we navigate the complexities of the digital age and beyond, the lessons learned from the past, particularly in the realm of cybersecurity and intelligence protection, will serve as invaluable guideposts in shaping a future where information flows freely, knowledge empowers, and human intelligence continues to evolve and adapt in an ever-changing world.

The history of intercommunal intelligence is interwoven with the challenges of protecting sensitive information and preventing its compromise. From the coded messages of ancient spies to the sophisticated encryption algorithms of the digital age, societies have continually sought to safeguard their knowledge and prevent its manipulation or exploitation by adversaries. The rise of the internet and the proliferation of digital technologies have brought about an unprecedented level of interconnectedness, facilitating the rapid exchange of information and ideas on a global scale. However, this interconnectedness also presents new vulnerabilities, as cyberattacks, misinformation campaigns, and surveillance efforts threaten the integrity of information and the trust upon which inter-communal intelligence is built.

In this dynamic landscape, the lessons of the past become even more crucial. The ancient art of cryptography, the development of secure communication channels, and the ongoing efforts to combat misinformation and propaganda provide valuable insights for navigating the challenges of the digital age. By learning from the successes and failures of past generations, we can develop strategies to protect our digital infrastructure, secure our communication networks, and foster a culture of critical engagement with information.

The future of intercommunal intelligence hinges on our ability to harness the empowering potential of technology while remaining vigilant in safeguarding digital freedoms and promoting a culture of open knowledge. By embracing cybersecurity principles, fostering media literacy, and empowering individuals and communities to protect their digital autonomy, we can shape a future where information flows freely, knowledge empowers, and human intelligence continues to evolve and adapt in an ever-changing world.

In the early days of civilization, intelligence gathering relied heavily on human actors: the cunning of spies, deciphering coded messages by skilled cryptographers, and carefully interpreting often unreliable reports from informants. Information security depended on messengers' trustworthiness, the strength of encryption techniques, and the ability to discern truth from deception. Intelligence hacking, in those times, might have involved intercepting messengers, bribing confidantes, or employing skilled codebreakers to unravel the secrets hidden within encrypted communications.

The invention of the printing press marked a turning point in the history of information dissemination. While it accelerated the spread of knowledge and empowered individuals with access to a broader range of ideas, it also introduced new challenges for intelligence protection. The mass production of printed materials made it easier for information to fall into the wrong hands, and the control of information became a central concern for those in power. Censorship, propaganda, and the suppression of dissenting voices emerged as tools to manipulate public opinion and maintain control over the narrative.

The development of electronic communication technologies, such as the telegraph and the telephone, further accelerated the pace of information flow, connecting communities across vast distances and blurring the lines between private and public communication. The security challenges escalated as new vulnerabilities emerged. The interception and decryption of electronic signals became a focus of intelligence agencies, and the potential for espionage and sabotage grew exponentially.

In the modern digital age, the proliferation of interconnected devices and the rise of the internet have created an unprecedented information explosion. While this hyper-connectivity has fostered collaboration, innovation, and cultural exchange on a global scale, it has also brought forth a new era of cybersecurity threats. The sheer volume and velocity of data flowing across networks make it increasingly difficult to protect sensitive information and ensure the integrity of communication channels.

Cyberattacks, ranging from data breaches and ransomware attacks to sophisticated disinformation campaigns and the manipulation of critical infrastructure, pose a significant threat to individuals, organizations, and national security. The compromise of intelligence in this hyper-connected world can have far-reaching consequences, eroding trust, disrupting essential services, and undermining the very foundations of societal stability.

Protecting information and preserving trust is paramount as we navigate this complex, ever-evolving digital landscape. The future of intercommunal intelligence hinges on our ability to develop robust cybersecurity measures, foster a culture of digital literacy, and remain vigilant against the evolving tactics of those who seek to exploit vulnerabilities and manipulate information for their gain.

The advent of electronic communication, with the telegraph's crackling Morse code messages and the telephone's disembodied voices traversing continents, irrevocably accelerated the pace of intercommunal intelligence. The world, once a patchwork of isolated communities, was rapidly knitting itself together through a network of invisible threads, connecting individuals and societies in unprecedented ways. News, ideas, and strategic information flowed across borders quickly, transforming diplomacy, commerce, and warfare.

However, this newfound interconnectedness came at a price. The technologies that enabled this unprecedented exchange of information also introduced new vulnerabilities. The ethereal nature of electronic communication made it susceptible to interception, eavesdropping, and manipulation by those seeking to exploit these vulnerabilities for their gain. The challenge of securing these communication channels and protecting sensitive information from prying eyes grew exponentially more complex.

Governments and militaries invested heavily in cryptography and codebreaking, engaging in a constant arms race to secure their communications while attempting to decipher the secrets of their adversaries. The emergence of radio communication further amplified these challenges, as wireless signals could be intercepted by anyone with the right equipment, blurring the lines between public and private communication.

The rise of electronic communication marked a pivotal moment in the history of intelligence, ushering in an era where the security and integrity of information became paramount. The challenges of protecting sensitive data from interception and manipulation grew increasingly complex, laying the groundwork for today's cybersecurity landscape.

Today, in the hyper-connected digital age, the flow of information has become a torrential flood, surging across traditional boundaries and forging a global village where communities intertwine in ways never imagined. The ease with which information is shared and knowledge disseminated at lightning speed has fostered collaboration, innovation, and cultural exchange on an unprecedented global scale. However, this interconnectedness comes at a cost. The challenges of misinformation, cyberattacks, and the erosion of privacy have become increasingly acute, casting a shadow over the bright promise of the digital revolution.

The technologies that facilitate the free flow of information also create vulnerabilities that malicious actors can exploit. Cyberattacks, from data breaches and ransomware attacks to sophisticated disinformation campaigns, threaten individuals, organizations, and national security. The ease with which misinformation can spread through social media and online networks poses a significant challenge to societal trust and cohesion. As personal data is collected, analyzed, and monetized by tech giants and governments, the erosion of privacy raises profound ethical concerns about autonomy and individual freedoms.

In this hyper-connected world, the integrity of information and digital infrastructure security are paramount. Cybersecurity measures, from robust firewalls and encryption protocols to sophisticated threat detection systems, are essential to protect sensitive data and maintain the trust that underpins online interactions. Digital literacy and critical thinking skills are crucial for individuals to navigate the complex digital landscape, discern truth from falsehood, and resist manipulation.

The challenges of misinformation, cyberattacks, and the erosion of privacy demand a collective response. Governments, organizations, and individuals must work together to create a more secure and resilient digital ecosystem. This includes investing in cybersecurity research and development, promoting education and awareness initiatives, and fostering international cooperation to combat cybercrime and protect digital freedoms.

The hyper-connected digital age presents both immense opportunities and unprecedented challenges. By embracing the empowering potential of technology while remaining vigilant in safeguarding digital freedoms and fostering a culture of critical engagement, we can navigate this complex landscape and ensure that the digital revolution catalyzes progress, innovation, and a more equitable and interconnected world.

The extreme difficulty of understanding the compromise of intelligence in today's interconnected world presents a formidable challenge to individuals, organizations, and nations alike. The technologies that connect us, facilitating the rapid exchange of information and ideas, also expose us to new and evolving vulnerabilities, making it increasingly difficult to discern the origins, intentions, and trustworthiness of the information that floods our digital landscapes.

In this era of hyper-connectivity, where the lines between the physical and digital worlds blur, the threat of intelligence compromise looms large. Cyberattacks, misinformation campaigns, and sophisticated social engineering tactics can readily infiltrate our networks, manipulate our perceptions, and undermine the foundations of trust upon which our societies are built.

The ease with which information can be shared and disseminated in the digital age amplifies the potential impact of compromised intelligence. A single fabricated news story, a cleverly disguised phishing email, or a deepfake video can spread like wildfire across social media platforms and online communities, sowing discord, eroding trust, and manipulating public opinion.

The challenge of understanding intelligence compromise is further compounded by the increasing sophistication of cyberattacks and the evolving tactics employed by malicious actors. Nation-state-sponsored hackers, cybercriminal organizations, and lone-wolf actors can leverage advanced technologies and exploit vulnerabilities to infiltrate systems, steal sensitive data, and disrupt critical infrastructure.

The interconnectedness of our digital world means that a cyberattack launched in one corner of the globe can have ripple effects across continents, impacting individuals, businesses, and governments alike. The NotPetya ransomware attack of 2017, for example, crippled businesses and critical infrastructure worldwide, causing billions of dollars in damages and highlighting the interconnectedness of our digital ecosystems.

In this environment, the ability to discern information's origins, intentions, and trustworthiness becomes paramount. Critical thinking, media literacy, and a healthy dose of skepticism are essential in navigating the digital landscape and safeguarding against manipulation and deception.

The challenge of understanding intelligence compromise demands a multifaceted approach. Individuals must be empowered with the knowledge and skills to identify and resist cyber threats, while organizations and governments must invest in robust cybersecurity infrastructure and proactive defense strategies. International cooperation and information sharing are crucial in combating the global nature of cyber threats and fostering a more secure and resilient digital world.

The journey of societal intercommunal intelligence is a testament to the human desire for connection, understanding, and progress. It is a narrative woven through time, tracing the evolution of how we share knowledge and build collective wisdom. As we navigate the complexities of the digital age and the uncharted territories beyond, the lessons etched in our history serve as guideposts, illuminating the path towards a future where information flows freely, knowledge empowers, and human intelligence continues its remarkable journey of evolution and adaptation.

Nevertheless, this journey is not without its perils. While offering unprecedented opportunities for connection and collaboration, the digital age also presents new challenges to information integrity and knowledge sharing security. The rise of cyber threats, from sophisticated hacking techniques to the spread of misinformation and propaganda, casts a shadow over the landscape of intercommunal intelligence.

The ease with which information can be manipulated, distorted, and weaponized in the digital realm demands a heightened awareness of cybersecurity and a commitment to protecting the integrity of knowledge. As we increasingly rely on digital technologies for communication, collaboration, and information sharing, safeguarding these channels from malicious actors becomes paramount.

The lessons learned from the past, from the ancient art of cryptography to the modern development of encryption technologies, remind us that protecting information has always been essential for advancing human societies. In today's interconnected world, where the boundaries between communities blur and information traverses the globe instantly, the security of our digital infrastructure is crucial for maintaining trust, fostering collaboration, and preserving the foundations of societal intelligence.

Cybersecurity challenges will undoubtedly grow more complex as we venture into the future. The rise of artificial intelligence, the emergence of virtual worlds, and the increasing integration of technology into our lives will create new vulnerabilities and demand innovative solutions.

However, the enduring human desire for connection, understanding, and progress will continue to drive us forward. By embracing the lessons of the past, fostering a culture of cybersecurity awareness, and investing in the development of robust security measures, we can ensure that the journey of societal intercommunal intelligence continues to illuminate the path toward a more informed, interconnected, and resilient future.

4 AI Mimicry and Impressioning

The Adversarial Deepfake Threat

The human fascination with understanding and controlling society has deep historical roots, stretching back to ancient civilizations and their intricate governance and social order systems. This chapter explores the evolution of societal observation and control, tracing its trajectory from the physical manifestations of power in the form of miniature cities and panopticons to the sophisticated digital surveillance systems that permeate the modern era.

From the intricate miniature cities crafted by societal elites, meticulously designed to mirror and monitor the lives of their subjects, to the imposing architecture of panopticons, where the watchful gaze of authority extended to every corner of society, the desire to observe, analyze, and regulate human behavior has been a persistent theme throughout history. Though rudimentary compared to today's digital surveillance technologies, these early forms of societal control laid the groundwork for the complex systems of monitoring and manipulation that shape our modern world.

The rise of technology has dramatically amplified the capacity for societal observation and control. The proliferation of digital devices, the interconnectedness of the internet, and the advent of artificial intelligence (AI) have created an unprecedented ability to gather, analyze, and utilize data on an unprecedented scale. This data-driven revolution has transformed the way we live, work, and interact with each other, blurring the lines between public and private spheres and raising profound questions about the future of individual autonomy and privacy.

As we navigate this complex landscape, the lessons learned from the past can illuminate the challenges and opportunities. By understanding the historical roots of societal observation and control, we can gain a deeper appreciation for the delicate balance between security and freedom, the benefits of technology, and the potential for its misuse. This chapter delves into this intricate interplay, exploring the evolution of surveillance, the rise of digital control, and the enduring human quest for autonomy and self-determination in an increasingly interconnected world.

In the opulent courts and bustling workshops of the Renaissance and Enlightenment periods, a fascination with intricate clockwork mechanisms took hold, captivating the imaginations of artisans, philosophers, and the elite alike. Skilled artisans, patronized by wealthy patrons and enlightened monarchs, poured their ingenuity into crafting miniature working models of societies – intricate microcosms of human activity, complete with whirring gears, delicately balanced levers, and lifelike automata that

DOI: 10.1201/9781003641506-4

mimicked the daily routines of citizens. These clockwork cities marvels of engineering and artistry, served not only as objects of wonder and entertainment but also as tools for understanding and, perhaps more importantly, controlling the complex dynamics of society.

These miniature worlds, encased in glass or meticulously crafted wooden frames, were more than mere toys or displays of technical prowess. They represented a deep-seated desire to comprehend the intricate workings of human society to reduce its complexities to a series of predictable mechanical interactions. The rhythmic movements of the automata, the synchronized interplay of gears and levers, and the carefully orchestrated cycles of day and night within these miniature cities reflected a yearning for order and predictability in a world often perceived as chaotic and unpredictable.

The fascination with clockwork mechanisms extended beyond entertainment and aesthetics. Philosophers and social theorists saw these intricate machines as a metaphor for society's functioning. The idea that human behavior could be understood and even predicted through the lens of mechanical principles gained traction, influencing the development of social sciences and the quest for a rational understanding of human interactions.

However, the allure of clockwork societies also carried a darker undercurrent – the desire for control. The ability to manipulate the mechanisms of these miniature worlds, to set the pace of life and dictate the actions of its inhabitants, hinted at a more profound ambition to control the real-world society. In its intricate perfection, the clockwork city became a symbol of both the human desire for understanding and the potential for manipulation and control.

By observing the miniature world in motion within their clockwork cities, elites could identify patterns, predict behaviors, and experiment with different forms of social engineering. The intricate mechanisms of the clockwork city mirrored the complexities of society, allowing rulers and social engineers to study the interplay of individual actions and collective dynamics. They observed how environmental changes, such as resource availability or social interactions, could influence behavior and trigger predictable responses. This knowledge empowered them to manipulate the levers of power, fine-tuning the city's mechanisms to achieve desired outcomes: increased productivity, social conformity, or the suppression of dissent.

The clockwork city became a microcosm of society, a controlled environment where the elites could observe, analyze, and manipulate social forces with a degree of detachment. They could experiment with different forms of social engineering, observing the effects of rewards, punishments, and social pressures on the miniature population. The insights gained from these experiments were then applied to the real world, shaping policies, social norms, and even architectural designs to influence behavior and maintain control.

Although limited in scope and technological sophistication, this early form of societal surveillance foreshadowed the sophisticated surveillance technologies that would emerge centuries later. The underlying principle remains: observing and analyzing human behavior to predict and control individuals and populations. The clockwork city, with its intricate mechanisms and miniature inhabitants, serves as a potent metaphor for the surveillance societies of today, where digital technologies and vast data collection networks enable unprecedented monitoring and manipulation.

The clockwork cities of the past, with their intricate mechanisms and meticulously engineered systems, find a modern echo in the ambitious android project. However, where those historical cities sought to control their 'inhabitants' physical movement and interactions, the modern iteration seeks to capture and analyze the digital essence of society itself. A comprehensive analog of society is meticulously captured and recorded through a vast, interconnected network of mobile devices, surveillance systems, and biometric control systems. This digital tapestry, woven from the threads of human activity, emotions, and interactions, provides a rich and fertile ground for analysis, manipulation, and the potential for unprecedented control.

Imagine a city where every movement, every conversation, every transaction is meticulously tracked and analyzed. The data streams from countless sensors, cameras, and digital devices converge to create a dynamic map of human behavior, revealing patterns, predicting trends, and exposing vulnerabilities. This intricate web of information, once the domain of science fiction, is rapidly becoming a reality, raising profound questions about privacy, autonomy, and the very nature of human society in the digital age.

The clockwork cities of the past, with their rigid structures and imposed order, may seem a far cry from the seemingly boundless freedom of the digital world. Nevertheless, beneath the surface of this digital utopia lies a potential for control that surpasses the wildest dreams of past autocrats. The ability to monitor, analyze, and manipulate the digital traces of human activity creates unprecedented opportunities for surveillance, social engineering, and the subtle shaping of behavior.

The Android project, with its ambition to create a digital mirror of society, stands as a stark reminder of the double-edged nature of technology. While it offers the potential for deeper understanding, enhanced efficiency, and improved social well-being, it also carries the risk of unprecedented control, manipulation, and the erosion of individual freedoms. As we navigate this complex landscape, the choices we make about the development and deployment of technology will determine the fate of human society in the digital age.

The android project, with its capacity to observe, learn, and adapt to the intricate dance of human social dynamics, mirrors the ambitions of those early visionaries who dreamt of clockwork cities – intricate mechanisms designed to regulate and orchestrate the flow of urban life. However, the goal of this ambitious project extends far beyond mere observation. It seeks to delve into the heart of human behavior, understand its motivations, predict its patterns, and, ultimately, influence its course on a societal scale.

This quest to unlock the secrets of human behavior through the lens of artificial intelligence raises profound ethical questions that reverberate through the corridors of power, the chambers of academia, and the hearts of individuals. The specter of privacy violation looms as the android's unblinking gaze captures the nuances of human interaction, recording conversations, analyzing facial expressions, and mapping the intricate networks of social connections. The notion of individual autonomy is challenged, as the android's ability to predict and influence behavior blurs the lines between free will and technological manipulation.

The potential for abuse of power is undeniable. In the hands of a malevolent actor, the android's insights into human behavior could be weaponized to sow discord, manipulate public opinion, or even subvert the democratic process. This project's ethical tightrope is precarious, demanding careful consideration of the potential consequences and a steadfast commitment to safeguarding human dignity and freedom.

In its ambition and complexity, the android project mirrors the human story, a narrative woven with threads of innovation, ambition, and the pursuit of knowledge, yet also shadowed by the potential for hubris, overreach, and the erosion of fundamental values. As we venture further into this uncharted territory, the ethical implications of the android project must be at the forefront of our minds, guiding our decisions and ensuring that the pursuit of knowledge serves the betterment of humanity, not its subjugation.

Today's surveillance systems, though vastly more sophisticated than their mechanical predecessors, share a common goal: to observe, analyze, and ultimately control human behavior. The panopticon concept, a prison design where inmates are constantly visible to an unseen observer, has found its digital equivalent in the ubiquitous surveillance technologies that permeate modern society. This digital panopticon, woven into the fabric of our cities, our online interactions, and even our devices, casts a long shadow over individual privacy and freedom.

From the watchful eyes of CCTV cameras perched on street corners to the sophisticated algorithms of facial recognition software that can identify individuals in a crowd, our physical movements are increasingly subject to scrutiny and analysis. The rise of smart cities, with their interconnected networks of sensors and data collection points, further amplifies this surveillance, tracking our commutes, monitoring our energy consumption, and even mapping our social interactions.

The digital realm offers no escape from this pervasive gaze. Our online activity is meticulously tracked and analyzed, from the websites we visit to the messages we exchange. Social media platforms, search engines, and online retailers collect vast personal data, constructing detailed profiles of our preferences, habits, and innermost thoughts and desires.

Often aggregated and anonymized, this data is used for targeted advertising, personalized recommendations, and even predictive policing. While these applications may offer certain conveniences and benefits, they come at a cost – the erosion of privacy and the potential for manipulation and control.

The implications of this digital panopticon are profound. The constant awareness of being watched can have a chilling effect on freedom of expression, leading to self-censorship and conformity. The potential for misuse of personal data by governments, corporations, or malicious actors poses a significant threat to individual autonomy and civil liberties.

As we navigate this increasingly surveilled world, the need to safeguard privacy and protect against the encroachment of the digital panopticon becomes paramount. This requires technological solutions, such as encryption and anonymization tools, and a societal shift toward greater awareness of surveillance practices and a renewed commitment to the values of individual freedom and autonomy.

As technology continues its relentless march forward, the tools of societal control are poised to become more sophisticated and insidiously pervasive. The rise of artificial intelligence, with its ever-growing capacity for machine learning and predictive analytics, casts a long shadow on the future of individual autonomy. This shadow hints at a world where human behavior is not only passively observed but actively predicted and manipulated with increasing accuracy.

Imagine a society where AI-powered surveillance systems can analyze vast troves of data, discerning patterns and predicting behaviors with chilling precision, where algorithms can identify potential dissenters before they even act, flagging them for preemptive intervention or targeted propaganda campaigns. The subtle nudges of personalized recommendations and targeted advertising can subtly shape opinions, influence choices, and steer individuals toward desired outcomes.

The notion of free will and independent thought could be eroded in such an environment, as the lines between genuine choice and technologically orchestrated influence become increasingly blurred. The potential for abuse is immense, raising profound ethical questions about the future of human agency in a world where technology increasingly shapes our thoughts, beliefs, and actions.

The challenge for future societies lies in navigating the complex and often fraught relationship between technological advancement and individual liberties. While technology offers immense potential for progress, innovation, and the betterment of humanity, it also carries the potential for misuse, manipulation, and the erosion of fundamental freedoms. The ethical implications of surveillance technologies, the potential for abuse of power by governments and corporations, and the erosion of privacy demand careful consideration and proactive measures to safeguard human autonomy in the digital age.

The increasing pervasiveness of surveillance technologies, from facial recognition systems to data tracking and social media monitoring, raises profound questions about the boundaries of privacy and the balance between security and individual liberties. While these technologies can undoubtedly be valuable tools for law enforcement and public safety, their potential for misuse and their chilling effect on freedom of expression and association cannot be ignored.

The challenge for future societies is to harness the benefits of technology while ensuring that its deployment aligns with our core values and respects the fundamental rights of individuals. This requires a nuanced approach that recognizes the importance of transparency and accountability in developing and implementing surveillance technologies. It demands a robust legal framework that protects against abuse of power and ensures that surveillance practices are subject to appropriate oversight and regulation.

Furthermore, fostering a culture of digital literacy and critical engagement with technology is crucial. By empowering individuals with the knowledge and skills to understand the implications of surveillance technologies, protect their digital privacy, and engage in informed discussions about the ethical implications of technological advancements, we can create a society that is both technologically advanced and deeply committed to the preservation of human autonomy.

In essence, the future of human societies hinges on our ability to navigate the complex interplay between technology and individual freedoms. By embracing the values of transparency, accountability, and ethical innovation, we can harness the transformative power of technology while safeguarding the fundamental liberties that define our humanity.

The journey from clockwork cities to digital shadows is a testament to the enduring human fascination with understanding and controlling the intricate mechanisms of society. From the panoptic gaze of Jeremy Bentham's architectural vision to the pervasive surveillance technologies of the modern era, the tools and techniques for observing, analyzing, and influencing human behavior have undergone a dramatic transformation. Nevertheless, beneath this technological evolution lies a persistent pursuit of knowledge and control, a quest to decipher the patterns of human behavior and shape the trajectory of societies.

In the clockwork cities of the Industrial Revolution, surveillance was often physical and localized, confined to the watchful eyes of factory supervisors, the meticulous record-keeping of bureaucracies, and the imposing presence of law enforcement. However, the rise of digital technologies has ushered in an era of unprecedented surveillance capabilities, where data collection, analysis, and dissemination transcend geographical boundaries and permeate the very fabric of our lives.

The digital shadows we cast in this hyper-connected world are woven from the intricate tapestry of our online activities, social media interactions, financial transactions, and even physical movements tracked by GPS and surveillance cameras. This wealth of data, when harnessed responsibly, can offer valuable insights into societal trends, public health patterns, and the effectiveness of policies and interventions. However, the potential for misuse and abuse of this information is a growing concern, raising ethical questions about privacy, autonomy, and the balance between security and individual freedoms.

As we navigate the complexities of the digital age, it is crucial to remain vigilant in protecting individual liberties, fostering a culture of ethical technology use, and ensuring that the pursuit of knowledge and understanding does not come at the cost of human autonomy and dignity. The seductive allure of control, amplified by the power of technology, must be tempered by a deep respect for human rights, a commitment to transparency and accountability, and a recognition that the accurate measure of a society lies not in its ability to control its citizens but in its capacity to empower them.

THE DANCE OF DECEPTION: UNVEILING THE INTENT OF CONTROL AND THE RISE OF COUNTER-IMPRESSIONING

In the age of information, where knowledge is power, and the flow of ideas shapes the destiny of nations, the battle for control often unfolds in the subtle realm of censorship and surveillance. Governments and powerful institutions, driven to maintain their grip on power, employ tactics to shape narratives, suppress dissent, and engineer conformity. This chapter delves into the intricate world of societal controls, exploring how censorship and surveillance can inadvertently reveal the hidden agendas of those in power. By analyzing these control mechanisms, individuals and communities can decipher the intent of the controlling authorities and design countermeasures, engaging in a complex dance of deception that has profound implications for cybersecurity.

Censorship, the suppression of information deemed undesirable or threatening, can take many forms, from outright banning books and websites to more subtle manipulation of media narratives and silencing dissenting voices. Surveillance, the covert observation and monitoring of individuals and communities, further extends the reach of control, enabling the collection of vast amounts of data that can be used to track, predict, and even manipulate behavior.

The application of censorship and surveillance, while intended to maintain control, can paradoxically reveal the very intentions and insecurities of the controlling authorities. The topics deemed too sensitive for public consumption, the individuals targeted for surveillance, and the narratives suppressed from the mainstream media all offer clues to the hidden agendas and vulnerabilities of those in power.

By carefully analyzing these control mechanisms, individuals and communities can gain valuable insights into the motivations and fears of the controlling authorities. This knowledge can be used to design countermeasures, challenge censorship, circumvent surveillance, and reclaim control over the flow of information. The rise of encryption technologies, anonymization tools, and decentralized platforms allows individuals to protect their privacy, access unfiltered information, and engage in open dialogue, even in the face of censorship and surveillance.

This dynamic interplay between control and resistance, between surveillance and subversion, creates a complex dance of deception, where individuals and communities constantly struggle to protect their autonomy and shape their narratives. The stakes are high, as the consequences of this struggle have far-reaching implications for cybersecurity, the preservation of truth, and the future of open societies.

Censorship, a tool as old as human history itself, casts a wide net in its attempts to control the flow of information and mold public perception. It manifests in myriad forms, from the blatant banning of books and the silencing of dissenting voices to the more subtle manipulation of search engine results and the algorithmic curation of social media feeds. By limiting access to specific ideas, narratives, or perspectives that challenge the status quo, authorities seek to maintain a firm grip on the dominant discourse, ensuring that only approved narratives reach the public consciousness. This suppression of alternative viewpoints creates an intellectual echo chamber where conformity reigns and critical thinking is stifled.

Surveillance, the silent partner of censorship, operates in the shadows, casting a pervasive gaze over the lives of individuals and communities. It is a relentless data collector, silently observing our digital footprints, mapping our social connections, and building comprehensive profiles that can be used for social control, manipulation, or even persecution. The rise of sophisticated surveillance technologies, for example, facial recognition software that can identify individuals in crowds to data mining algorithms that sift through vast troves of personal information, has ushered in an era of unprecedented scrutiny, where the boundaries of privacy blur and the chilling effect of constant monitoring casts a pall over freedom of expression.

This insidious combination of censorship and surveillance creates a fertile ground for the erosion of trust, the suppression of dissent, and the manipulation of public opinion. When individuals are denied access to diverse perspectives, fear expressing their true beliefs, and constantly feel the watchful eye of authority, the foundations of a free and open society begin to crumble.

The combined effect of censorship and surveillance casts a long shadow over society, stifling the free flow of information and the essence of human expression and intellectual exploration. When individuals live under the constant threat of their words and actions being monitored, scrutinized, and potentially punished, a chilling effect takes hold, silencing dissenting voices and discouraging the open exchange of ideas.

Fear becomes a pervasive force, driving individuals to self-censor, suppress their true thoughts and opinions, and conform to the prevailing narrative, even if it contradicts their beliefs or values. The public sphere becomes a stage for performative compliance, where outward agreement masks inner dissent, and genuine dialogue is replaced by cautious tiptoeing around sensitive topics.

Critical thinking, the cornerstone of intellectual progress and innovation, withers under such conditions. The lack of exposure to diverse perspectives, the suppression of dissenting viewpoints, and the fear of challenging the status quo create an intellectual vacuum where conformity is rewarded and independent thought is discouraged. The consequences are far-reaching, hindering societal progress, stifling creativity, and impeding the ability to adapt to new challenges and opportunities.

This chilling effect extends beyond politics and social discourse, seeping into the fabric of cultural and artistic expression. Artists, writers, and musicians may hesitate to explore controversial themes or challenge societal norms, fearing reprisal or censorship. The result is an increasingly homogenized cultural landscape devoid of the vibrant diversity of thought and expression that fuels creativity and innovation.

The combined effect of censorship and surveillance is a form of societal self-sabotage, suppression of the qualities that drive progress, foster resilience, and enable societies to thrive in an ever-changing world. By silencing dissenting voices, discouraging critical thinking, and rewarding conformity, we create a society that is not only less free but also less intelligent, less adaptable, and ultimately less capable of navigating the complexities of the 21st century and beyond.

With its inherent yearning for freedom and autonomy, the human spirit rarely remains passive in the face of control. When confronted with censorship, surveillance, and attempts to limit self-expression, individuals and communities rise in defiance, devising ingenious strategies to circumvent restrictions and reclaim their digital freedoms. This dance of deception is a testament to human resilience, a dynamic interplay between the forces of control and the unwavering pursuit of liberty in the digital age.

It is a dance that demands a deep understanding of the tactics employed by controlling authorities. Censors may attempt to suppress information, manipulate narratives, and restrict access to knowledge, but resilient individuals and communities find ways to counter these maneuvers. They develop tools and techniques to bypass censorship, utilizing proxy servers, virtual private networks (VPNs), and encrypted communication channels to access blocked websites and share information freely.

The dance of deception also involves challenging the ever-present gaze of surveillance technologies. From facial recognition software to data tracking algorithms, surveillance systems seek to monitor and control individual behavior. However, resilient communities fight back, employing tactics to obfuscate their identities and movements. They do literally and figuratively wear masks to shield themselves from the prying eyes of surveillance cameras and data-hungry algorithms. They utilize anti-tracking tools and privacy-enhancing technologies to reclaim control over their digital footprints and protect their personal information.

This dance is not merely a technical arms race; it is a cultural tug-of-war, a battle for the hearts and minds of individuals. It is a struggle to preserve the values of free expression, open inquiry, and individual autonomy in the face of forces that seek to homogenize, control, and suppress.

The dance of deception is a testament to the enduring human spirit, a beacon of hope in the face of adversity. It is a reminder that pursuing freedom and self-expression finds a way to flourish even in the most restrictive environments. It is a dance that will continue to evolve as technology advances and the struggle for control intensifies, but it is a dance that humanity is determined to lead, ensuring that the digital age remains a space where individual liberties and cultural diversity can thrive.

One powerful strategy of counter-impression is leveraging technology to bypass censorship and access information that authorities seek to suppress. VPNs, anonymization tools, and encrypted communication channels act as digital shields, allowing individuals to circumvent restrictions and access a free and open internet. These tools empower citizens to bypass government firewalls, access blocked websites and social media platforms, and connect with others worldwide, fostering a global exchange of ideas and perspectives.

The development of alternative platforms and decentralized technologies further challenges the control of centralized authorities. These platforms, operating independently of government or corporate oversight, provide spaces for free expression, unfiltered information sharing, and the organization of collective action. They represent a digital rebellion against censorship and surveillance, empowering individuals and communities to reclaim control over their digital lives and narratives.

Another form of counter-impression involves actively manipulating the data collected by surveillance systems. By intentionally generating false or misleading information, individuals can disrupt the surveillance apparatus, sow confusion, and challenge the data's accuracy. While potentially risky, this tactic can be a powerful form of resistance, highlighting the limitations of surveillance technologies and the resilience of the human spirit in the face of control.

Imagine individuals wearing clothing designed to confuse facial recognition systems or artists creating subversive installations that generate false positives for surveillance algorithms. These creative acts of resistance challenge the efficacy of surveillance technologies and serve as powerful expressions of defiance, reminding us that human ingenuity and the desire for freedom can find ways to subvert even the most sophisticated systems of control.

The ongoing battle between control and resistance, between the forces of censorship and the unwavering pursuit of open knowledge, has profound implications for the cybersecurity landscape. As the tactics of control grow increasingly sophisticated, employing advanced surveillance technologies, AI-powered monitoring systems, and subtle information manipulation, resistance forces become more creative and determined to develop countermeasures. This dynamic interplay creates a constantly evolving landscape of vulnerabilities and exploits, a digital arms race where both sides seek to gain an advantage in the ongoing struggle for information dominance.

One manifestation of this struggle is the rise of adversarial attacks against AI-powered surveillance systems. Recognizing the increasing reliance on AI for monitoring and control, individuals and communities are developing techniques to deceive and manipulate these systems, challenging their accuracy and undermining their effectiveness. By understanding the vulnerabilities of AI models and exploiting their inherent limitations, those seeking to resist control can turn the very tools of surveillance against their creators.

Another form of counter-impression involves actively manipulating the data collected by surveillance systems. By intentionally generating false or misleading information, individuals can disrupt the surveillance apparatus, create confusion, and challenge the data's accuracy. While potentially risky, this tactic can serve as a powerful form of resistance, highlighting the limitations of surveillance technologies and the resilience of the human spirit in the face of control.

This dynamic interplay between control and resistance, between surveillance and subversion, is a defining characteristic of the digital age. As technology evolves at an unprecedented pace, the battle for information dominance will intensify, demanding constant vigilance and adaptation from both sides. The societies that thrive in this environment will embrace the values of open knowledge, critical thinking, and individual autonomy while also recognizing the importance of cybersecurity and the need to protect against the evolving tactics of control and manipulation.

The risk of adversarial attacks amplifies within a society stifled by censorship and control. This heightened risk stems from external actors seeking to exploit vulnerabilities and internal actors who may be motivated to disrupt or manipulate systems for political or ideological purposes.

The technologies designed to protect against cyber threats can become double-edged swords, capable of being wielded as tools of oppression in the hands of controlling authorities. Surveillance systems, firewalls, and content filtering mechanisms, while intended to safeguard networks and maintain order, can be readily repurposed to monitor citizens, suppress dissent, and restrict access to information.

Furthermore, the efforts of individuals and communities to circumvent censorship, while crucial for preserving freedom of expression and accessing unfiltered knowledge, can inadvertently create new vulnerabilities that malicious actors can exploit. Using anonymization tools, circumvention software, and decentralized platforms while empowering individuals to bypass restrictions can also obscure identities and create opportunities for malicious actors to operate undetected.

This complex interplay between control, resistance, and vulnerability underscores the delicate balance between security and freedom in the digital age. While censorship and control may create an illusion of stability, they often foster an environment of distrust, paranoia, and vulnerability to manipulation. The suppression of information and the stifling of dissent can hinder the development of a robust cybersecurity culture, leaving individuals and communities ill-equipped to defend against the evolving tactics of cyber adversaries.

In contrast, societies that foster open knowledge, encourage critical thinking, and empower individuals to protect their digital freedoms are better equipped to navigate the complex landscape of cyber threats. By promoting transparency, accountability, and a culture of cybersecurity awareness, these societies can harness the empowering potential of technology while mitigating the risks of adversarial attacks and manipulation.

The dance of deception played out in the shadowy realm of censorship, surveillance, and cultural resistance is a defining characteristic of our digital age. It is a complex and ever-evolving ballet where power dynamics shift, technologies intertwine with human aspirations, and the pursuit of knowledge clashes with the forces of control. In this intricate dance, the societies that thrive will be those that not only embrace the empowering potential of technology but also remain vigilant in safeguarding digital freedoms and fostering a culture of critical engagement with information.

These societies will be the ones that recognize the transformative power of technology while acknowledging its potential for misuse and manipulation. They will champion cybersecurity awareness, educating citizens about the evolving landscape of cyber threats and empowering them with the knowledge and skills to navigate the digital world safely and responsibly. They will invest in robust cybersecurity infrastructure, protecting critical systems and sensitive data from malicious actors while ensuring the privacy and security of their citizens.

Furthermore, these societies will cultivate a culture of open knowledge and critical engagement, recognizing that the free flow of information, the diversity of perspectives, and the ability to challenge conventional thinking are essential for progress and innovation. They will resist the temptation to censor ideas, suppress dissent, or manipulate narratives, understanding that such actions ultimately stifle creativity, hinder progress, and erode trust.

By understanding the tactics of control, developing effective countermeasures, and promoting a culture of cybersecurity awareness, we can navigate this complex landscape and ensure that the pursuit of knowledge and preserving individual liberties remain at the forefront of human progress. This requires a multifaceted approach encompassing technological innovation, educational initiatives, and a collective commitment to safeguarding the values underpinning a free and open society.

We must invest in developing privacy-enhancing technologies, encryption tools, and secure communication channels that empower individuals to protect their digital identities and resist surveillance. We must promote digital literacy and critical thinking skills, equipping citizens to discern truth from falsehood in the vast ocean of online information. We must foster a culture of transparency and accountability, holding governments and corporations responsible for their technology use and impact on individual liberties and societal well-being.

In this ongoing dance of deception, the future of human societies hinges on our ability to harness the empowering potential of technology while remaining vigilant in safeguarding the values that define our humanity. By embracing the principles of openness, critical engagement, and a steadfast commitment to digital freedoms, we can ensure that pursuing knowledge and preserving individual liberties remain the guiding stars of human progress in the digital age and beyond.

THE ART OF COUNTER-IMPRESSION: SOCIETY'S STRATEGIES FOR RESISTING CONTROL

Throughout history, the human spirit has demonstrated a remarkable capacity for resistance, persistently devising ingenious strategies to counter attempts at control and preserve autonomy in the face of power imbalances. This chapter delves into the fascinating realm of counter-impressions, where individuals and communities engage in a dynamic dance of defiance, employing subtle and overt tactics to challenge surveillance, circumvent censorship, and reclaim control over their narratives and identities.

This inherent drive to resist control is rooted in the essence of human nature, a deep-seated yearning for freedom of thought, expression, and self-determination. Whether facing the oppressive weight of authoritarian regimes, the subtle manipulation of social norms, or the pervasive gaze of technological surveillance, individuals and communities have consistently found ways to subvert, challenge, and redefine the boundaries of power.

Counter-impression manifests in many forms, from the clandestine operations of spies and the coded messages of dissidents to the artistic expressions of rebellion and the everyday acts of defiance against conformity. It is a testament to human ingenuity, adaptability, and the unwavering pursuit of freedom in a world where power dynamics are constantly in flux.

This chapter will explore the diverse landscape of counter-impressioning, tracing its historical roots, examining its manifestations in the digital age, and considering its implications for the future of individual autonomy and societal resilience. From the ancient art of espionage to the modern tactics of digital resistance, we will uncover the strategies individuals and communities employ to challenge surveillance, circumvent censorship, and reclaim control over their narratives and identities.

Espionage, one of the oldest and most enduring forms of counter-impressions, has been a constant throughout human history, employed by states, empires, and organizations to gather intelligence, disrupt enemy operations, and influence political outcomes. Operating in the shadows, cloaked in secrecy and deception, spies have long relied on disguise, infiltration, and psychological manipulation to gain access to restricted information and subvert the control of adversaries.

In the ancient world, espionage was already a well-honed craft. With its vast territories and complex political landscape, the Roman Empire employed a network of spies and informants to gather intelligence on rival factions, potential rebellions, and external threats. These spies, often recruited from the lower classes or even from among enslaved people, would infiltrate enemy camps, gather information on troop movements, and report back to their Roman handlers.

During the Cold War, espionage reached new heights of sophistication, as the ideological clash between the United States and the Soviet Union fueled a global intelligence war. Both superpowers invested heavily in espionage networks, deploying agents to infiltrate each other's governments, military, and scientific institutions. The stakes were high, as intelligence gathered through espionage could sway the balance of power and determine the outcome of conflicts.

The effectiveness of espionage lies in its ability to circumvent censorship, bypass traditional security measures, and exploit human vulnerabilities. Spies, by adopting false identities and weaving elaborate cover stories, can infiltrate enemy organizations and gain the trust of unsuspecting individuals. They can gather information that would otherwise be inaccessible, manipulate perceptions, spread disinformation, and even sabotage operations, undermining the control of those in power.

The tools and techniques of espionage have evolved from invisible ink and dead drops to sophisticated surveillance technologies and cyber espionage tactics today. However, the fundamental principles of espionage remain the same: the ability to gather information, manipulate perceptions, and influence outcomes through secrecy, deception, and the exploitation of human trust.

In the digital age, the timeless dance of counter-impression has evolved, taking on new forms as individuals and communities leverage the technologies meant to control them. With its decentralized architecture and global reach, the internet has become a formidable tool for digital resistance, a virtual battlefield where the forces of surveillance and censorship clash with the unyielding human desire for freedom and autonomy.

The rise of encryption technologies, like VPNs and anonymization tools, empowers individuals to cloak their digital identities and circumvent the watchful eyes of censors. Armed with these digital shields, activists and dissidents in repressive regimes can communicate securely, organize protests, and disseminate information that would otherwise be suppressed. The internet becomes a lifeline, a conduit for truth and resistance in the face of oppressive forces.

The emergence of decentralized platforms, operating independently of centralized control, further challenges the dominance of traditional power structures. These platforms offer alternative spaces for expression, collaboration, and information sharing, fostering a sense of community and shared purpose among those who seek to resist censorship and reclaim their digital sovereignty.

The internet, once envisioned as a utopian space for the free exchange of ideas, has become a contested territory where the battle for control and freedom rages. In this dynamic landscape, counter-impression tactics continue to evolve, adapting to new forms of surveillance and censorship. From the use of adversarial AI to generate synthetic media that challenge the authenticity of information to the development of blockchain-based platforms that resist censorship and promote transparency, the digital age has ushered in a new era of resistance, where the power of technology is harnessed to defend the fundamental human rights of freedom of expression and access to information.

Encryption technologies, VPNs, and anonymization tools have become a cornerstone of digital self-defene in an increasing surveillance and censorship era. These technologies empower individuals to protect their digital privacy, cloak their online activities from prying eyes, and circumvent the restrictions imposed by repressive regimes. Activists and dissidents operating under the shadow of authoritarian governments rely on these tools to communicate securely, organize protests, and disseminate information that would otherwise be suppressed. Encryption scrambles their messages, making them unreadable to eavesdroppers, while VPNs mask their location and identity, allowing them to bypass government firewalls and access blocked websites. Anonymization tools further enhance their privacy, making tracing their online activities and identities more difficult.

The rise of decentralized platforms and blockchain technology represents another front in the struggle for digital freedom. These technologies challenge the control of centralized authorities, offering alternative spaces for expression, collaboration, and information sharing that operate independently of government or corporate oversight. Decentralized social media platforms, for example, allow users to connect and communicate without fear of censorship or surveillance, while blockchain-based file-sharing systems enable the secure and anonymous distribution of information. These technologies represent a powerful countermeasure to the increasing centralization of power in the digital realm, empowering individuals and communities to reclaim control over their digital lives and resist the encroachment of censorship and surveillance.

Cultural resiliency, the inherent human capacity to adapt and resist in the face of encroaching control, often manifests in subtle yet powerful forms of counter-impression. In an era of pervasive surveillance, where the watchful eyes of technology seem to permeate every aspect of our lives, individuals and communities are reclaiming their autonomy and challenging the omnipresent gaze of the digital panopticon. These acts of resistance, while often understated, represent a profound assertion of human agency in a world increasingly defined by technological surveillance.

Consider the simple act of covering one's face, a seemingly mundane gesture that takes on new meaning in the context of facial recognition technologies. A mask, a scarf, or even strategically placed hair can disrupt the algorithms that seek to identify and categorize us, reclaiming a degree of anonymity in public spaces and challenging the notion of constant surveillance as a societal norm.

Similarly, the obfuscation of license plates, whether through specialized covers or creative DIY solutions, subverts the automated tracking and monitoring of individual movements. These seemingly small acts of defiance chip away at the surveillance infrastructure, reminding us that we are not merely data points to be tracked and analyzed but autonomous individuals with the right to privacy and freedom of movement.

Beyond these physical acts of counter-surveillance, cultural resiliency also manifests in the digital realm. The development and adoption of privacy-enhancing technologies, such as encryption tools and anonymization software, empower individuals to shield their online activities from prying eyes and protect their digital identities. The rise of decentralized platforms and alternative communication channels challenges the dominance of centralized social media giants, offering spaces for uncensored expression and the free exchange of information.

While often overlooked or dismissed as insignificant, these subtle forms of counter-impressions represent a powerful undercurrent of resistance against the encroachment of surveillance technologies and the erosion of privacy. They are a testament to the enduring human desire for autonomy, self-expression, and preserving individual identity in a world increasingly defined by technological control.

Covering one's face, whether with a mask, a scarf, or even strategically placed hair, becomes a subtle yet powerful resistance in a world increasingly permeated by surveillance technologies. It disrupts the gaze of facial recognition systems, those digital eyes that seek to categorize and catalogue our identities, reclaiming a degree of anonymity in public spaces and challenging the erosion of privacy in the digital age.

Similarly, obscuring license plates or adopting anti-surveillance clothing thwarts attempts to track and monitor individuals' movements, creating a sense of agency and control in a world where our every move is often recorded and analyzed. These seemingly small acts of defiance carry a profound message: that individuals have the right to control their information, to move freely without being tracked, and to resist the encroachment of technologies that erode our fundamental rights to privacy and autonomy.

These counter-surveillance measures are not merely about evading detection; they represent a broader cultural shift and a growing awareness of the importance of privacy in the digital age. They challenge the normalization of constant surveillance, the insidious creep of technologies that seek to monitor and control our every move. They serve as a potent reminder that we are not mere data points to be collected and analyzed but individuals with the right to define our identities and navigate the world on our terms.

In a society where technology is increasingly used to shape and control our behavior, these acts of resistance become even more crucial. They represent a reclaiming of agency, a refusal to be passively monitored and manipulated. They embody the spirit of cultural resilience, the inherent human desire to protect our autonomy and safeguard the fundamental freedoms that define our humanity.

THE DOUBLE-EDGED SWORD OF TECHNOLOGY

In its inherently dualistic nature, technology presents both a promise of liberation and a threat of control. While offering tools for individuals and communities to resist surveillance and censorship, it simultaneously empowers those seeking to enhance these mechanisms. The development of sophisticated AI-powered surveillance systems, capable of analyzing vast troves of data and discerning intricate patterns of behavior, poses a formidable challenge to cultural resiliency and preserving individual liberties.

These AI-driven systems, fueled by the exponential growth of data and the increasing sophistication of machine learning algorithms, can monitor online activities, track physical movements, and even analyze emotional expressions with unprecedented precision. This pervasive surveillance capability threatens to erode privacy, stifle dissent, and create an environment where individuals feel constantly scrutinized and manipulated.

The implications for cultural resiliency are profound. As surveillance technologies become more sophisticated and ubiquitous, the ability of individuals and communities to challenge authority, protect their privacy, and maintain control over their narratives becomes increasingly tricky. The chilling effect of constant surveillance can stifle creativity, discourage dissent, and lead to self-censorship, undermining the foundations of a vibrant and resilient culture.

Furthermore, using AI in surveillance systems raises concerns about bias and discrimination. If these systems are trained on biased data or programmed with flawed algorithms, they can perpetuate and even amplify societal inequalities. This could lead to the disproportionate targeting and surveillance of marginalized communities, further exacerbating social divisions and undermining trust in institutions.

The challenge lies in harnessing the beneficial applications of technology while mitigating its potential for misuse and abuse. This requires a multifaceted approach encompassing ethical considerations, robust regulations, and the development of counter-surveillance technologies and strategies.

By fostering a culture of transparency and accountability in developing and deploying AI-powered surveillance systems, we can ensure that these technologies serve the interests of society rather than undermining its fundamental freedoms. We can foster a more resilient and informed citizenry by promoting digital literacy and empowering individuals with the knowledge and tools to protect their privacy.

The ongoing struggle between surveillance, privacy, control, and freedom will continue to shape the digital landscape. By recognizing the dualistic nature of technology and actively engaging in AI's ethical development and deployment, we can strive toward a future where technology empowers individuals and communities rather than eroding their fundamental rights and freedoms.

Deepfake technology, with its uncanny ability to create realistic yet entirely fabricated videos and audio recordings, adds a troubling new dimension to the already complex landscape of information warfare. The power to manipulate perceptions and spread disinformation through deepfakes poses a significant threat to trust, eroding the foundation upon which informed decision-making and social cohesion rest.

Imagine a world where seeing is no longer believing, a political leader's speech can be seamlessly altered to fabricate inflammatory remarks, or a loved one's voice can be cloned to orchestrate a convincing scam. Deepfakes have the potential to sow discord, fuel conflict, and undermine democratic processes by blurring the lines between reality and fabrication.

However, amidst this challenge lies a glimmer of hope. The technologies that enable surveillance and manipulation can also be harnessed to counter these threats. Researchers are developing sophisticated anti-deepfake algorithms that can detect subtle inconsistencies and artifacts within fabricated media, exposing the deceptive hand behind their creation. These algorithms, coupled with the promotion of media literacy and critical thinking skills, can empower individuals to discern truth from falsehood, question the authenticity of information, and resist manipulated 'narratives' seductive power.

By fostering a society that values critical inquiry, embraces digital literacy, and champions the development of ethical AI countermeasures, we can navigate the treacherous terrain of deepfakes and safeguard the integrity of information. The battle against disinformation is not merely technological but a struggle to preserve truth, trust, and the foundations of human understanding.

The art of counter-impression is a dynamic dance between control and resistance, a timeless struggle between surveillance forces and the human yearning for autonomy and self-expression. Throughout history, individuals and communities have engaged in this intricate dance, devising ingenious strategies to challenge authority, circumvent censorship, and reclaim control over their narratives and identities. From the clandestine whispers of ancient spies to the defiant keystrokes of digital dissidents, the human spirit has consistently sought to subvert the gaze of power and assert its freedom in the face of control.

In the shadows of history, espionage has long served as a tool for the marginalized and the oppressed, a means of undermining power structures and gaining access to the corridors of influence. Spies, cloaked in secrecy and deception, have infiltrated the ranks of the powerful, gathering intelligence, disrupting operations, and challenging the narratives propagated by those in control. Their subversion, often hidden from public view, has shaped history, empowering the powerless and challenging the status quo.

In the modern era, the rise of digital technologies has ushered in new forms of counter-impressions, as individuals and communities leverage the power of the internet to resist surveillance, circumvent censorship, and reclaim their digital autonomy. Encryption technologies, anonymization tools, and decentralized platforms have become the weapons of choice in this digital struggle for freedom, allowing individuals to protect their privacy, communicate securely, and challenge the dominance of centralized authorities.

The cultural resilience demonstrated through acts of counter-surveillance, such as face coverings and license plate obfuscation, underscores the deep-seated human desire for autonomy and self-determination. These acts of defiance, while seemingly minor, represent a powerful assertion of individual agency in the face of pervasive surveillance technologies. They are a testament to the enduring human spirit, which refuses to be silenced or controlled, even in the face of overwhelming power.

The art of counter-impression is an ongoing dynamic, a constant negotiation between the forces of control and the human desire for freedom. It is a testament to the resilience, ingenuity, and unwavering pursuit of autonomy that define the human spirit. As technology continues to evolve, the landscape of counter-impression will undoubtedly transform, presenting new challenges and opportunities. However, the enduring human yearning for freedom, self-expression, and the right to shape one's narrative will continue to fuel the fires of resistance, ensuring that the dance between control and subversion continues for generations.

As technology continues its relentless march forward, the landscape of counter-impressions will inevitably undergo a profound transformation, presenting novel challenges and exciting opportunities for individuals and communities seeking to resist control and assert their autonomy. The tools that enable surveillance, censorship, and manipulation will also spark the development of new strategies and technologies designed to counter these forces, ensuring that the human spirit of resilience, ingenuity, and the pursuit of freedom remains undeterred.

The rise of artificial intelligence, while potentially enhancing the capabilities of those seeking to control and manipulate, will also empower individuals with tools to protect their privacy, circumvent censorship, and challenge dominant narratives. The development of advanced encryption techniques, anonymization tools, and decentralized platforms will continue to evolve, creating a dynamic interplay between control and resistance in the digital realm.

The struggle for control over information and identity will intensify as those in power seek to leverage technology to maintain their grip, while those seeking freedom and autonomy will harness the same technologies to resist and subvert. The battleground will shift, with new fronts emerging in virtual reality, augmented reality, and the metaverse, where the lines between the physical and digital worlds blur.

However, amidst this technological upheaval, the enduring human spirit of resilience, ingenuity, and the pursuit of freedom will remain constant. The desire for self-determination, the yearning for authentic expression, and the unwavering belief in the importance of individual liberties will continue to fuel the development of new strategies and technologies to counter control and reclaim agency.

This dynamic interplay between technology, control, and resistance will shape the future of counter-impression. It will be a future where individuals and communities leverage their collective intelligence, creativity, and technological prowess to challenge the forces that seek to limit their freedoms and homogenize their identities.

Cultural diversity will flourish in the future, not despite technology but because of it. The digital realm will become a canvas for the expression of diverse voices, a platform for preserving cultural heritage, and a space for celebrating human differences.

Pursuing freedom, preserving cultural identity, and celebrating human diversity will continue to inspire acts of resistance, innovation, and the relentless pursuit of a future where technology empowers rather than enslaves.

THE 'ACHILLES' HEEL OF AI: VULNERABILITY IN FOOLED AND IMPRESSIONED DEEP MODELS

Artificial intelligence has woven itself into the fabric of modern life, permeating nearly every facet of our daily existence. From the mundane to the extraordinary, AI's influence is undeniable. It powers our smartphones, curating our digital experiences and connecting us to a global information and communication network. It guides our entertainment choices, recommending movies, music, and books tailored to our preferences. It even assists in critical domains like medical diagnoses, analyzing complex medical images, and aiding healthcare professionals in life-altering decisions. Furthermore, on the horizon, AI-powered autonomous vehicles promise to revolutionize transportation, offering the potential for safer and more efficient mobility.

Deep learning models, a subset of AI inspired by the human brain's structure and function, have achieved remarkable feats. These models, composed of intricate layers of interconnected nodes, have demonstrated superhuman performance in tasks that once seemed exclusive to human intelligence. They can recognize objects and faces accurately, surpassing human capabilities in specific visual recognition tasks. They can process and understand natural language, translate between languages, generate human-quality text, and even engage in conversations that blur the lines between human and machine interaction. Furthermore, in the realm of games, deep learning models have conquered complex strategic challenges, defeating world champions in games like Go and chess.

However, despite their impressive achievements, these robust AI systems are not infallible. They remain susceptible to vulnerabilities and weaknesses that malicious actors can exploit. This chapter delves into the perils of fooled and impressioned deep models, exploring how adversarial attacks can manipulate their inputs, exploit their biases, and ultimately compromise their integrity. By understanding these vulnerabilities, we can develop more robust and secure AI systems, ensuring that AI's transformative power is harnessed for humanity's betterment while mitigating its potential risks.

THE ILLUSION OF INTELLIGENCE: FOOLING DEEP MODELS

While capable of impressive pattern recognition and prediction feats, deep learning models operate under a veil of apparent intelligence. Their inner workings, however, are fundamentally rooted in complex mathematical operations and statistical analysis, far removed from the nuanced understanding and commonsense reasoning that characterize human thought. This inherent limitation renders them susceptible to adversarial attacks, a form of digital manipulation where subtle alterations to input data can lead to unexpected and often alarmingly erroneous outputs.

Imagine a sophisticated image recognition system meticulously trained on vast datasets to identify objects accurately. For instance, a seemingly innocuous image of a stop sign would be readily recognized and categorized. However, armed with knowledge of the model's inner workings, an attacker could introduce carefully crafted perturbations to the image, subtle shifts in pixel values imperceptible to the human eye yet capable of derailing the AI's perception. The stop sign, now subtly altered, might be misclassified as a speed limit sign, a yield sign, or even an entirely unrelated object, potentially leading to disastrous consequences in a real-world scenario.

This fooling of the deep model underscores its vulnerability to adversarial manipulation, exposing the inherent limitations of its artificial intelligence. While capable of impressive feats within its training domain, the deep learning model lacks the contextual awareness and common-sense reasoning that would allow it to recognize the absurdity of its misclassification. It operates within the confines of its mathematical framework, blind to the real-world implications of its errors.

This vulnerability to adversarial attacks highlights the crucial distinction between artificial intelligence and human intelligence. While AI excels in pattern recognition and prediction, it lacks the broader understanding, adaptability, and critical thinking skills that characterize human cognition. As we increasingly rely on AI systems for critical tasks, from medical diagnoses to autonomous driving, understanding and mitigating these vulnerabilities become paramount to ensuring safety, security, and the responsible development of artificial intelligence.

THE PERILS OF IMPRINTING: BIASES AND BACKDOORS

While capable of impressive pattern recognition and prediction feats, deep learning models are not immune to the biases and imperfections present pattern recognition and prediction feats. However, deep learning models are not immune to the biases and imperfections in the data they learn from. This inherent vulnerability can lead to the imprinting of unwanted biases, resulting in discriminatory or unfair outcomes that perpetuate societal inequalities. Imagine a facial recognition system trained on a dataset predominantly composed of images of white males. This system is likely to exhibit lower accuracy for individuals from other demographic groups, such as women or people of color, potentially leading to misidentification, wrongful arrests, or even denial of services.

Moreover, the vulnerability of deep learning models extends beyond unintentional biases. Malicious actors can intentionally inject backdoors into the training data, creating hidden vulnerabilities that can be exploited later. These backdoors act as secret triggers, causing the AI system to malfunction or produce specific outputs when presented with certain inputs. For instance, a backdoor implanted in a self-driving car's AI model could be designed to trigger a sudden braking maneuver when the car encounters a particular visual pattern, potentially causing a collision.

The implications of these vulnerabilities are far-reaching, impacting not only individual fairness and privacy but also societal trust in AI systems. As AI becomes increasingly integrated into critical infrastructure, healthcare, finance, and other essential domains, the consequences of biased or compromised AI models could be devastating.

Addressing these challenges requires a multi-pronged approach. First, ensuring diversity and inclusivity in the training data is crucial to mitigate the risk of unwanted biases. This involves actively collecting data representing the full spectrum of human experiences and demographics. Second, rigorous testing and validation of AI models are essential to identify and address potential biases or backdoors before deployment. This includes employing techniques like adversarial training and explainable AI to enhance the transparency and robustness of AI systems. Finally, fostering a culture of ethical AI development and deployment is paramount, emphasizing fairness, accountability, and transparency in the design and use of AI systems.

ADVERSARIAL ATTACKS: EXPLOITING THE 'ACHILLES' HEEL

Adversarial attacks represent a significant threat to the integrity and reliability of AI systems. These attacks, crafted maliciously, exploit AI 'models' inherent vulnerabilities by introducing carefully designed inputs that disrupt their normal functioning or manipulate their outputs. The consequences of such attacks can range from minor inconveniences to severe repercussions, including financial losses, privacy breaches, safety hazards, and even loss of life.

Evasion attacks, a typical adversarial attack, aim to bypass the AI system by subtly manipulating input data. For instance, an attacker could make imperceptible changes to an image, such as adding noise or altering pixels, causing an image recognition system to misclassify the object. Similarly, attackers can craft malicious emails or websites that evade spam filters or antivirus software by subtly altering their characteristics in cybersecurity.

Poisoning attacks, on the other hand, target the training process of AI models. By injecting malicious data into the training dataset, attackers can compromise the model's integrity or introduce backdoors that can be exploited later. This can lead to AI systems that produce biased or discriminatory outputs or behave erratically under specific conditions.

Model extraction attacks represent a more sophisticated threat, where attackers aim to steal the AI model. This allows them to replicate the model's functionality, potentially using it for malicious purposes or gaining insights into its vulnerabilities. The theft of intellectual property, the creation of counterfeit AI systems, and the development of more effective adversarial attacks are potential consequences of model extraction.

The vulnerability of AI systems to these attacks underscores the urgent need for robust security measures and ongoing research to develop more resilient AI models. This includes incorporating adversarial examples into the training process, implementing techniques to detect and filter out adversarial inputs, and designing AI models that are more transparent and explainable, making it easier to identify and mitigate vulnerabilities.

As AI continues to permeate critical aspects of our lives, from healthcare and finance to transportation and national security, the consequences of adversarial attacks could be devastating. By understanding the nature of these threats and investing in developing more secure and resilient AI systems, we can harness the transformative power of AI while safeguarding against its potential risks.

MITIGATING THE RISKS: TOWARD ROBUST AI

Researchers are diligently working on innovative techniques to bolster the resilience of AI models against the looming threat of adversarial attacks. These techniques represent a multi-pronged approach, addressing the vulnerabilities of AI from various angles and striving to create more robust and secure systems.

Adversarial training, a key technique in this endeavor, involves exposing the AI model to a barrage of adversarial examples during its training process. By confronting the model with these carefully crafted inputs designed to deceive, researchers aim to vaccinate the AI, enhancing its ability to recognize and resist such attacks in real-world scenarios. This approach, akin to a rigorous training regimen for a martial artist, prepares the AI to anticipate and counter the deceptive tactics of adversaries.

Defensive distillation, another promising technique, employs a secondary model as a smoothing filter for the primary AI model's output. This secondary model, trained on the original model's predictions, learns to generalize and "smooth out" the decision boundaries, making the AI less susceptible to minor perturbations in the input data. This approach can be likened to adding a layer of shock absorption to a vehicle, reducing the impact of unexpected bumps and jolts.

Input validation and sanitization techniques focus on fortifying the AI's defenses at the entry point. By implementing rigorous checks and filters, researchers aim to detect and neutralize adversarial inputs before they can wreak havoc on AI decision-making. This approach can be compared to a security checkpoint at a building entrance, carefully screening individuals and preventing unauthorized access.

Explainable AI (XAI) represents a paradigm shift in AI development, emphasizing transparency and interpretability. By designing AI models that can provide insights into their decision-making process, researchers aim to demystify the black box of AI, making it easier to identify vulnerabilities and potential biases. XAI can be likened to a transparent machine, allowing us to peer into its inner workings and understand the rationale behind its decisions.

While still in their early stages of development, these techniques represent a crucial step toward creating more robust and secure AI systems. As AI continues to permeate critical aspects of our lives, from healthcare and finance to transportation and national security, safeguarding these systems from adversarial attacks cannot be overstated. By investing in research, fostering collaboration between experts, and promoting responsible AI development, we can harness the transformative power of AI while mitigating its potential risks.

The vulnerability of fooled and impressioned deep models to adversarial attacks highlights the urgent need for ongoing research and development to enhance the security and robustness of AI systems. As AI continues to permeate critical aspects of our lives, from healthcare and finance to transportation and national security, the consequences of adversarial attacks could be devastating. These attacks, ranging from subtle manipulations of input data to more complex exploitations of model vulnerabilities, can lead to erroneous outputs, discriminatory outcomes, and even safety hazards. By understanding the vulnerabilities of AI models and developing effective mitigation strategies, we can harness the transformative power of AI while safeguarding against its potential risks.

The intricate dance between control and resistance, surveillance and subversion, has been a constant undercurrent throughout human history, shaping our societies, technologies, and understanding of freedom. This chapter has ventured into the fascinating realm of counter-impressions, where individuals and societies alike have risen to defy authority, protect their autonomy, and reclaim control over their narratives in the face of those who seek to dominate and define.

From the ancient art of espionage, with its cloak-and-dagger maneuvers and whispers in the shadows, to the modern tactics of digital resistance, where encryption and anonymity tools become weapons against surveillance, the human spirit has consistently demonstrated an unyielding resolve to challenge the boundaries of control and assert its inherent freedom. This drive to resist, subvert, and redefine is woven into the very fabric of human nature, a testament to our enduring yearning for autonomy and self-determination.

We have explored the clandestine operations of spies, the subtle acts of defiance against surveillance technologies, and the collective movements that rise to challenge oppressive regimes. Each example is a powerful reminder that the pursuit of control is often met with an equally powerful force of resistance. The human spirit, it seems, refuses to be confined by the boundaries imposed by those who seek to dominate.

This dynamic interplay between control and resistance, between surveillance and subversion, is not merely a historical curiosity but an ongoing struggle that continues to shape our world today. In the digital age, where technology has become both a tool of empowerment and a means of control, the battle for autonomy and self-expression is being waged on new fronts. The rise of mass surveillance, the erosion of privacy, and the attempts to manipulate online narratives have sparked a resurgence of resistance, with individuals and communities employing creative and innovative tactics to reclaim their digital freedoms.

This ongoing dance between control and resistance is a testament to the enduring human spirit, our unwavering pursuit of freedom, and our refusal to be defined by the forces that seek to confine us. It is a dance that will continue to shape human history as we navigate the complexities of a world where technology empowers and threatens, connects and isolates, and ultimately challenges us to define the very meaning of freedom in the digital age.

The examples explored in this chapter, spanning the clandestine operations of spies to the subtle acts of defiance against surveillance technologies, vividly illustrate counter-impression's diverse and ever-evolving nature. With its reliance on deception, infiltration, and the art of assuming false identities, espionage has long served as a tool for undermining power structures and gaining access to coveted information. From the shadowy figures of ancient history to the sophisticated operatives of the modern era, spies have navigated the treacherous terrains of secrecy, risking their lives to gather intelligence, disrupt enemy operations, and influence the course of history.

In the digital age, the rise of encryption, anonymization tools, and decentralized platforms has empowered individuals and communities to resist control, circumvent censorship, and challenge the dominance of centralized authorities. These technological tools have become the modern arsenal of counter-impressioning, enabling citizens to protect their privacy, access unfiltered information, and organize collective action in ways never imagined. With its decentralized nature and global reach, the internet has become a fertile ground for digital resistance, where individuals can reclaim their autonomy and challenge the encroachment of surveillance and control.

The cultural resilience demonstrated through acts of counter-surveillance, such as face coverings and license plate obfuscation, underscores a deep-seated human desire for autonomy and self-determination. These acts of defiance, while seemingly small and perhaps even trivial to some, represent a powerful assertion of individual agency in the face of increasingly pervasive surveillance technologies. They are a quiet rebellion against the erosion of privacy, reclaiming control over one's identity and movements in a world where technology threatens to render individuals transparent and subject to constant monitoring.

This resistance is not merely a rejection of technology itself but a pushback against its unchecked deployment in ways that infringe upon fundamental freedoms. It is a recognition that technology should serve humanity, not vice versa. Obscuring one's face or license plate is a symbolic act, reclaiming anonymity in a society where every move is tracked, analyzed, and categorized. It means saying, "I am not just data points; I am a human being with the right to privacy and self-determination."

This cultural resilience speaks to the enduring human spirit, the inherent drive to resist control and assert individuality. It is a reminder that despite seemingly overwhelming technological power, the human desire for freedom and autonomy remains a potent force. When multiplied across individuals and communities, these small acts of defiance can coalesce into a powerful wave of resistance, shaping the trajectory of technology and ensuring that it serves to empower rather than enslave.

The relationship between technology and control is indeed a complex and ever-shifting dance, a dynamic interplay where the tools that empower individuals to resist surveillance can, paradoxically, be wielded to enhance it. This inherent duality of technology presents a profound challenge to cultural resiliency and the preservation of truth in the digital age.

On the one hand, technology has undeniably become a powerful instrument for individuals and communities seeking to challenge authority, circumvent censorship, and protect their digital freedoms. Encryption technologies, anonymization tools, and decentralized platforms offer a means to safeguard privacy, facilitate secure communication, and foster the free exchange of information even in the face of repressive regimes and surveillance apparatuses.

However, the rapid advancement of technology also brings with it the potential for unprecedented levels of surveillance and control. The development of sophisticated AI-powered surveillance systems, capable of analyzing vast troves of data and identifying patterns of behavior with chilling accuracy, poses a formidable challenge to individual autonomy and privacy preservation.

Deepfake technologies, with their ability to generate realistic yet fabricated videos and audio recordings, further blur the lines between reality and artificiality, casting a shadow of doubt over the authenticity of information and eroding trust in digital media. The potential for deepfakes to be weaponized for propaganda, disinformation, and the manipulation of public opinion underscores the urgent need for critical media literacy and the development of robust countermeasures.

In this dynamic landscape, the relationship between technology and control is in constant flux, a delicate balance between empowerment and manipulation, liberation and oppression. The societies that navigate this complexity successfully will embrace the empowering potential of technology while remaining vigilant in safeguarding digital freedoms and fostering a culture of critical engagement with information.

Preserving truth, protecting privacy, and cultivating trust in a technologically mediated world will require a collective effort, a commitment to ethical innovation, and a steadfast defence of the values underpinning a free and open society.

In the ever-shifting digital landscape, where the boundaries between the physical and virtual realms blur and the lines between truth and falsehood become increasingly obscured, the ability to adapt, innovate, and critically engage with technology emerges as a paramount skill for individuals and societies alike. The future of counter-impressions, the art of resisting control and preserving autonomy in the face of surveillance and manipulation, hinges on our capacity to harness the empowering potential of technology while remaining vigilant in safeguarding digital freedoms and fostering a culture of media literacy and critical thinking.

This requires a multifaceted approach that embraces the transformative power of technology while acknowledging its potential for misuse and the erosion of privacy. It demands a critical understanding of the digital landscape and awareness of how technology can shape perceptions, manipulate behavior, and control narratives.

The future of counter-impression lies in our ability to adapt to the ever-evolving tactics of surveillance and control, to innovate new strategies for resistance, and to critically engage with the information that bombards us from all directions. It lies in our ability to discern truth from falsehood, to question the narratives presented to us, and to seek out diverse perspectives and alternative sources of information.

This requires a collective effort and a commitment to fostering a culture of media literacy and critical thinking, where individuals are empowered to question, analyze, and evaluate the information they encounter. It requires recognition that technology is not a neutral tool but a powerful force that can be wielded for both good and ill.

By embracing the principles of adaptability, innovation, and critical engagement, we can navigate the complex digital landscape and ensure that technology serves as a tool for empowerment, liberation, and the preservation of human autonomy. We can reclaim control over our narratives, challenge the forces of surveillance and manipulation, and shape a future where the digital realm fosters freedom of expression, diversity of thought, and the flourishing of human potential.

The ongoing struggle between control and freedom, between surveillance and subversion, is a dynamic woven into the very fabric of human history, and it will undoubtedly continue to shape the trajectory of societies in the digital age and beyond. This tension between the desire for order and the yearning for autonomy, between the impulse to restrict and the urge to express, is fundamental to the human condition. The societies that thrive in this dynamic environment, navigating the complexities of technological advancement and the ever-present challenges to individual liberty, will embrace individual autonomy, open knowledge, and critical engagement with information.

These societies will recognize that progress and innovation flourish in an atmosphere where individuals can think critically, question authority, and express their perspectives without fear of reprisal. They will understand that the suppression of ideas, the censorship of information, and the stifling of dissent ultimately lead to intellectual stagnation, social fragmentation, and a vulnerability to manipulation.

By fostering a culture of resilience, these societies will equip their citizens with the tools and mindset to navigate the digital landscape's complexities, discern truth from falsehood, and resist the encroachment of surveillance and control. They will cultivate a spirit of innovation, encouraging the development of technologies that empower individuals, protect privacy, and promote the free flow of information.

With a steadfast commitment to digital freedoms, these societies will champion the principles of open access, net neutrality, and the right to privacy in the digital realm. They will recognize that these freedoms are not merely abstract ideals but essential pillars of a just and equitable society, where technology is a tool for empowerment rather than a means of control.

In essence, the societies that thrive in the digital age will embrace the transformative potential of technology while remaining vigilant in safeguarding the values that define our humanity. By fostering a culture of resilience, innovation, and a steadfast commitment to digital freedoms, we can ensure that the human spirit continues to defy control, reclaim agency, and shape a future where technology empowers rather than enslaves, where knowledge liberates rather than confines, and where the flame of individual liberty burns brightly in the digital landscape.

5 Can We Hack Humanity?
The Cybersecurity Threat to Intelligence in a Boxed-Up Society

The human race stands on the precipice of its most daring endeavor yet: sending humans to Mars. This unprecedented journey presents a crucible of challenges unlike anything ever experienced. From the prolonged weightlessness of a seven-month voyage to the bone-jarring gravitational forces upon landing, the human body and mind will be pushed to their limits. The Martian environment, with its alien day-night cycles and the isolating distance from loved ones, will further test the resilience of our astronauts.

In this extreme environment, personal and inter-societal human intelligence becomes unstable. The cognitive effects of prolonged isolation, the stress of constant heart-racing activity, and the disruption of circadian rhythms create a perfect storm for psychological vulnerability. To cope, humans on Mars must rely on artificial intelligence for support, from managing daily routines to executing complex mission assignments.

This reliance on AI introduces a new layer of complexity. As humans interact with sophisticated AI systems, there is a risk of "model imprinting," where the AI's behavior and decision-making patterns begin to influence humans. This could lead to unforeseen consequences, potentially compromising the mission's objectives.

This chapter delves into the uncharted territory of human intelligence in the face of Martian extremes. We explore the psychological and social challenges, the ethical implications of relying on AI, and the potential risks of model imprinting. Understanding these challenges is not just an academic exercise; it is crucial for ensuring the success and safety of future Martian pioneers. The insights gained from this exploration will be vital in preparing humanity for its most ambitious journey yet.

Both human colonies on Mars and early primitive human societies on Earth share the fundamental challenge of survival in a harsh and unforgiving environment. They both grapple with limited resources, requiring innovative approaches to shelter, food production, and waste management. Social cohesion and cooperation are paramount for both, as the community's success hinges on individuals working together to overcome adversity. Technological innovation, albeit at different levels of sophistication, plays a crucial role in both scenarios, enabling adaptation and progress. Moreover, both Martian colonies and early human societies need to establish governance structures and social norms to maintain order

DOI: 10.1201/9781003641506-5

and ensure the well-being of their members. Finally, both share the psychological challenges of isolation, confinement, and the need to adapt to an unfamiliar environment, demanding psychological resilience and a strong sense of community to thrive.

These challenges include limited resources, harsh environmental conditions, and potential social conflicts. The need for advanced technology and specialized skills to sustain life on Mars may create a significant knowledge and experience gap compared to early humans, who relied on essential tools and survival instincts. This disparity could lead to unforeseen vulnerabilities and difficulties adapting to unexpected situations. The psychological impact of isolation and confinement in a Martian habitat could also exacerbate these challenges, potentially hindering the colony's ability to thrive and maintain social cohesion.

Establishing the first human habitat on Mars marks a significant milestone in human history and presents unique challenges for developing inter-societal intelligence and cybersecurity. The isolated and confined nature of the Martian habitat, coupled with the reliance on advanced technology for survival, could lead to the development of specialized skills and knowledge within Martian society, potentially creating an intelligence gap between Mars and Earth. This gap could pose cybersecurity risks, as a Martian society might develop unique vulnerabilities not readily apparent to Earth-based security systems and protocols.

The experience of operating the International Space Station (ISS) has provided valuable insights into the challenges of maintaining cybersecurity in an isolated, technologically advanced environment. The ISS cybersecurity framework, which includes data protection, network security, and incident response measures, can be a foundation for developing robust cybersecurity protocols for Martian habitats. However, the unique characteristics of the Martian environment, such as the distance from Earth and the reliance on autonomous systems, will necessitate further adaptations and innovations in cybersecurity practices.

The interaction between the Martian habitat and Earth-based ground stations will be crucial for maintaining cybersecurity. Regular communication and data exchange will be necessary to update security protocols, share threat intelligence, and coordinate incident response efforts. However, the distance between Mars and Earth could introduce latency and communication challenges, potentially hindering real-time cybersecurity monitoring and response. Therefore, developing robust and resilient communication systems and protocols for autonomous cybersecurity measures on Mars will be essential.

The cultural and social dynamics within the Martian habitat could also impact cybersecurity. The close-knit community and shared experiences of the Martian settlers could foster a strong sense of trust, potentially leading to a relaxed approach to cybersecurity within the habitat. However, this trust could be exploited through social engineering tactics, highlighting the need for ongoing cybersecurity awareness training and education within Martian society.

Overall, developing inter-societies intelligence and cybersecurity for the first human habitat on Mars presents unique challenges and opportunities. By learning from the experiences of the ISS and other space endeavors and proactively addressing the vulnerabilities associated with the Martian environment and social dynamics, we can establish a secure and resilient foundation for human life on Mars.

A boxed-up community with minimal societal intelligence interaction could severely compromise the security of the first human colonies on Mars.

Reduced Individual and Collective Intelligence: Isolation and limited social interaction can hinder cognitive function, adaptability, and the development of crucial social and problem-solving skills. This "intelligence gap" can make individuals more vulnerable to manipulation, internal conflicts, and errors in judgment, compromising security in a high-risk environment like Mars.

Social interaction and exposure to diverse perspectives are crucial in cognitive development and acquiring essential life skills. Engaging in social interactions can enhance cognitive flexibility, adaptability, and the ability to navigate complex social situations. These skills are vital for maintaining individual well-being and promoting a secure and harmonious community, particularly in a challenging and isolated environment like a Mars colony. Prioritizing the development of social intelligence alongside technical skills can help mitigate these risks and foster a more resilient and adaptable community. This could involve incorporating diverse social interaction opportunities, fostering a strong sense of community, and implementing training programs that enhance critical thinking, psychological resilience, and awareness of social engineering tactics.

Increased Susceptibility to Social Engineering Attacks: A decline in social intelligence could make colonists more vulnerable to psychological manipulation and social engineering tactics. In a confined and isolated setting, the consequences of such attacks could be severe, potentially leading to breaches in security protocols, sabotage, or the compromise of critical systems. These tactics exploit psychological vulnerabilities and trust to access sensitive information or systems. In such a setting, the consequences of these attacks could be severe. Attackers could gain unauthorized access to critical systems, disrupt operations, or compromise life support infrastructure. Social engineering could also instigate conflict or discord among colonists, further destabilizing the community and hindering their ability to function effectively. The lack of external support and the challenges of a harsh environment could exacerbate the impact of such attacks, making it crucial to prioritize social intelligence training and awareness to mitigate these risks.

Impaired Group Cohesion and Decision-Making: Social interaction is crucial for fostering trust, cooperation, and effective decision-making within a group. A lack of social intelligence could lead to internal conflicts, impaired communication, and poor judgment, hindering the colony's ability to respond effectively to emergencies or security threats. Social interaction is the bedrock for establishing trust, fostering cooperation, and cultivating effective decision-making within any group, especially in a Mars colony's challenging and isolated environment. A rich tapestry of social interactions allows individuals to understand each other's perspectives, build rapport, and develop shared norms and values. This foundation of trust is essential for effective teamwork, conflict resolution, and collective problem-solving.

Conversely, a lack of social intelligence, characterized by difficulties perceiving and interpreting social cues, navigating social dynamics, and responding appropriately in social situations, can significantly hinder the colony's ability to function effectively. This deficiency can lead to miscommunications, misunderstandings, and interpersonal conflicts, eroding trust and cooperation among colonists. Furthermore, impaired social intelligence can impede effective decision-making, particularly in high-pressure situations. Individuals lacking social intelligence may struggle to accurately assess the perspectives and motivations of others, leading to misinterpretations and flawed judgments. This can hinder the colony's ability to respond swiftly and decisively to emergencies, technical malfunctions, or even potential security threats, jeopardizing the safety and well-being of the entire colony. In the high-stakes environment of a Mars colony, where survival depends on the ability to work together cohesively and adapt to unforeseen challenges, fostering social intelligence is not merely a desirable trait but a critical necessity.

Erosion of Cultural Resilience: Cultural resilience is essential for adapting to unforeseen challenges and maintaining morale in a demanding environment like Mars. A boxed-up community with limited social interaction could experience a decline in cultural resilience, making them less equipped to handle crises and external threats, including those that might arise from social engineering attacks. Cultural resilience, the ability of a community to adapt, persevere, and maintain its core values and identity in the face of challenges, is essential for the success of a Mars colony. Social interaction fosters a sense of shared identity, collective purpose, and the development of crucial social skills. It allows for the exchange of diverse perspectives, facilitating creative problem-solving and adaptability to unforeseen challenges. A decline in social interaction could weaken these essential qualities, making the colony more susceptible to social engineering tactics that exploit psychological vulnerabilities and erode trust.

Furthermore, cultural resilience is crucial for maintaining morale and psychological well-being in a Mars colony's demanding and isolated environment. A strong sense of community and shared cultural identity can buffer against stress, isolation, and potential conflict. A decline in cultural resilience could lead to a decline in mental and emotional well-being, making colonists more vulnerable to manipulation and less equipped to handle the challenges of establishing a new society on Mars. To foster cultural resilience, a Mars colony should prioritize social interaction, encourage diverse cultural expression, and promote a sense of shared purpose and collective identity. This could involve creating opportunities for social gatherings, supporting cultural events, and implementing educational programs that enhance social awareness and critical thinking skills. By cultivating

cultural resilience, a Mars colony can enhance its ability to adapt, persevere, and thrive in the face of challenges, ensuring its inhabitants' long-
term success and well-being.

Mitigating the risks associated with a boxed-up community on Mars requires
a multifaceted approach that prioritizes the development of social intelligence and
technical skills. This could involve fostering a strong sense of community through
regular social events, shared meals, and collaborative projects. Encouraging participation in decision-making processes and conflict-resolution workshops could further
enhance social intelligence. Training programs that focus on critical thinking, emotional regulation, and recognizing social engineering tactics could empower colonists to identify and respond to potential threats. Additionally, it could be beneficial
to incorporate virtual reality simulations or interactive scenarios that expose colonists to diverse social situations and challenge their social reasoning skills. Future
Mars colonies can cultivate a more resilient and secure environment for all inhabitants by promoting social interaction, fostering community spirit, and enhancing
social intelligence.

PREAMBLE: SECURITY IN THE MARTIAN FRONTIER

Establishing a human colony on Mars represents an unknown leap, a testament to
human ingenuity and our relentless pursuit of exploration. However, this endeavor
also presents unique challenges and necessitates a comprehensive approach to security that extends beyond traditional paradigms. The Martian environment is unforgiving, and the colony's survival depends on the integrity of its infrastructure, its
inhabitants' well-being, and its social fabric's resilience.

This security framework acknowledges the extraordinary challenges of establishing a human presence on Mars. It recognizes that security is not merely the absence
of threats but a proactive and holistic endeavor encompassing physical safety, psychological well-being, social cohesion, and critical infrastructure protection.

GUIDING PRINCIPLES:

At the core of this framework are three fundamental principles:

1. **Safeguarding Human Life:** The paramount concern is the preservation of
   human life. All security measures must prioritize the physical and psychological safety of every colonist. This includes protection from environmental hazards, accidents, and potential internal or external threats that could
   jeopardize their well-being.
2. **Protecting Critical Infrastructure:** The colony's survival hinges on the
   integrity and functionality of its critical infrastructure. This encompasses
   life support systems, power generation, communication networks, habitat
   structures, and all essential resources. Security measures must ensure these
   systems' continuous operation and resilience against potential accidental or
   malicious disruptions.

3. **Preserving Colony Integrity:** The colony's success depends on maintaining a cohesive and resilient social structure. This includes a sense of community, promoting cooperation, and establishing clear protocols for conflict resolution. Security measures must address potential social vulnerabilities and ensure the colony's social fabric remains strong in facing challenges.

Security on Mars is a shared responsibility. Every colonist plays a vital role in maintaining a safe and secure environment. This requires open communication, mutual respect, and a willingness to collaborate in the face of challenges. The colony's security framework must foster a culture of collective responsibility, where every individual is empowered to ensure the safety and well-being of the community.

This framework is not merely a set of rules but a guiding philosophy that underscores the interconnectedness of security, technology, and human behavior in the unique context of a Martian colony. It is a living document that will evolve alongside the colony, adapting to new challenges and incorporating lessons learned. Embracing principles of the culture of collaboration, the Mars colony can thrive as a testament to human resilience, ingenuity, and our capacity to overcome extraordinary challenges.

Physical Security and Habitat Integrity

The physical security and integrity of Martian habitats are paramount for ensuring the safety and well-being of colonists. Measures must be implemented to safeguard against the unique challenges posed by Mars's harsh environment, including:

- **Radiation Shielding:** Developing robust radiation shielding to protect inhabitants from harmful solar and cosmic radiation. This could involve incorporating radiation-resistant materials into habitat construction, creating underground shelters, or implementing artificial magnetic fields for deflection.
- **Thermal Regulation:** Maintaining a stable and comfortable temperature within habitats despite extreme external temperature fluctuations. This necessitates advanced insulation techniques, efficient heating and cooling systems, and Martian regolith's potential for thermal mass utilization.
- **Atmospheric Control:** Ensuring a breathable atmosphere within habitats and protecting against breaches or leaks. This involves robust airlocks, redundant oxygen generation systems, and advanced sensor networks for continuous atmospheric monitoring.
- **Structural Integrity:** Designing habitats to withstand the stresses of Martian conditions, including dust storms, meteorite impacts, and potential seismic activity. This necessitates using durable materials, innovative structural designs, and regular maintenance protocols.

Strict access control protocols are essential for preventing unauthorized entry into Martian habitats and critical infrastructure. This involves:

- **Multi-factor Authentication:** Implementing multi-factor authentication systems that combine biometrics, access cards, and/or passcodes to verify identity and grant access.
- **Surveillance and Monitoring:** Utilizing surveillance cameras, motion sensors, and intrusion detection systems to monitor access points and detect unauthorized entry attempts.
- **Secure Airlocks:** Designing airlocks with multiple security layers to prevent unauthorized passage and maintain atmospheric integrity.
- **Access Logging and Auditing:** Maintaining detailed logs of all access attempts and activities to track movement and identify potential security breaches.

EMERGENCY RESPONSE

Comprehensive emergency response plans are crucial for handling unforeseen events and ensuring the safety of Martian colonists. This includes:

- **Habitat Breach Protocols:** Develop clear procedures for responding to habitat breaches, including rapid sealing, atmospheric replenishment, and evacuation protocols if necessary.
- **Medical Emergency Response:** Equipping habitats with medical facilities and trained personnel to handle medical emergencies, including injuries, illnesses, and potential psychological challenges.
- **Equipment Malfunction Response:** Establishing procedures for addressing equipment malfunctions, including backup systems, repair protocols, and contingency plans for critical system failures.
- **Communication and Coordination:** Ensuring reliable communication systems and protocols for coordinating emergency response efforts and communicating with Earth-based support teams.

Protecting vital resources is essential for the survival and sustainability of Martian colonies. This involves:

- **Water Management:** Implementing water recycling and purification systems, safeguarding water sources from contamination, and developing strategies for efficient water usage.
- **Oxygen Generation:** Ensuring a continuous oxygen supply through redundant oxygen generation systems, regular maintenance, and backup reserves.
- **Food Production:** Developing sustainable food production systems, including hydroponics or Martian agriculture, and protecting food stores from contamination or spoilage.
- **Energy Management:** Utilizing renewable energy sources, such as solar or wind power, and implementing energy-efficient technologies to ensure a reliable power supply.
- **Security Measures:** Implementing security measures to prevent theft, sabotage, or unauthorized access to resource storage and distribution systems.

DIGITAL SECURITY AND NETWORK PROTECTION

Securing communication networks and data systems is paramount to the success and safety of any Mars colony. This involves implementing robust cybersecurity measures to protect against various threats, including unauthorized access, data breaches, and malicious attacks. Strong authentication protocols, encryption techniques, and intrusion detection systems are essential for safeguarding sensitive data and maintaining the integrity of critical systems. Regular security audits, vulnerability assessments, and penetration testing can help identify and address weaknesses in the digital infrastructure. Furthermore, establishing secure communication protocols and data backup systems is crucial to ensure continuity of operations in the event of a cyberattack.

CRITICAL INFRASTRUCTURE PROTECTION

Protecting critical infrastructure from cyber threats is of utmost importance in a Mars colony, where the survival and well-being of the inhabitants depend on the reliable operation of essential systems. This includes safeguarding power generation and distribution systems, life support infrastructure, communication networks, and other vital services. Implementing robust cybersecurity measures, such as intrusion detection systems, network segmentation, and access controls, is crucial to prevent disruptions and ensure the continuous operation of these critical systems. Regular security assessments and vulnerability patching are essential to maintain a strong security posture against evolving cyber threats.

DATA INTEGRITY

Maintaining the accuracy and reliability of scientific data, research findings, and operational records is crucial for advancing knowledge and the success of a Mars colony. Implementing data integrity measures, such as data validation checks, version control systems, and secure storage protocols, is essential to prevent data corruption or manipulation. Regular data backups and disaster recovery plans can help ensure the availability and recoverability of critical data in the event of system failures or cyberattacks. Furthermore, promoting a culture of data quality and integrity among researchers and personnel is vital to minimizing human error and maintaining the trustworthiness of scientific and operational data.

EXTERNAL COMMUNICATION SECURITY

Protecting communication channels with Earth is essential for a Mars colony to maintain contact with mission control, receive critical updates, and ensure the safety of its inhabitants. This involves securing communication links against interception, disruption, or manipulation by implementing robust encryption protocols and authentication mechanisms. Developing redundant communication systems and backup strategies help ensure continuous connectivity during technical failures or cyberattacks. Regularly monitoring and analyzing communication traffic for anomalies can help detect and respond to potential threats to external communication security.

PSYCHOLOGICAL SECURITY

Maintaining psychological well-being is paramount in a Mars colony's isolated and challenging environment. Comprehensive mental health support services should be readily available, including access to trained professionals, counselling resources, and strategies for stress management. These services should address the unique psychological challenges of isolation, confinement, and the demanding nature of space exploration. Establishing clear protocols for conflict resolution is essential to foster a harmonious social environment within the colony. This includes developing effective communication strategies, providing mediation resources, and implementing conflict resolution training programs.

Furthermore, proactive measures to mitigate stress and promote psychological resilience among colony members are crucial. This could involve incorporating mindfulness and relaxation techniques, creating opportunities for recreation and leisure activities, and fostering a supportive community environment. Building a strong sense of community and mutual support is vital for enhancing psychological security and overall well-being. Regular social events, shared meals, and collaborative projects can help strengthen social bonds and create a sense of belonging. Prioritizing psychological well-being alongside physical health and safety is essential for any future Mars colony's long-term success and sustainability.

ADAPTATION, EVOLUTION, AND CONTINUOUS IMPROVEMENT

Security protocols must be regularly reviewed and adapted to align with the colony's growth and evolution. As the colony expands, new challenges and vulnerabilities will emerge, necessitating adjustments to security measures. This includes incorporating feedback from colonists, analyzing security incidents, and conducting regular audits to identify areas for improvement. Establishing a culture of continuous improvement, where security protocols are constantly evaluated and refined, is essential for maintaining a secure environment.

TECHNOLOGICAL ADVANCEMENTS

Staying at the forefront of technological advancements is crucial for enhancing colony security. As new technologies emerge, they should be evaluated for their potential to improve security measures and address existing vulnerabilities. This could involve incorporating advanced surveillance systems, automated threat detection tools, or novel access control mechanisms. Embracing innovation and integrating cutting-edge technologies can significantly strengthen the colony's defenses against evolving threats.

RESEARCH AND DEVELOPMENT

Supporting research into Mars-specific security challenges is essential for developing innovative solutions. The unique environment of Mars presents distinct security considerations that require dedicated research efforts. This could involve investigating the impact of Martian conditions on security equipment, developing specialized countermeasures for Mars-specific threats, or exploring the potential of Martian resources for enhancing security. Fostering a culture of research and development within the colony can lead to groundbreaking solutions that ensure long-term security and resilience. The challenges early-life habitat societies face on Mars, encompassing both primitive and advanced aspects, will likely induce significant cultural shifts. The need to adapt to a harsh and unfamiliar environment, coupled with the reliance on advanced technologies for survival, may lead to a reprioritization of values and social norms. The traditional emphasis on individualism and competition might give way to a greater appreciation for collaboration, community spirit, and resourcefulness. Additionally, the experience of living in a confined and isolated habitat could foster a heightened sense of interdependence and a re-evaluation of priorities, potentially leading to a more minimalist and sustainable lifestyle.

The advanced challenges specific to Mars, such as the need for complex life support systems, radiation protection, and closed-loop resource management, could drive a cultural shift towards greater scientific understanding and technological proficiency. The reliance on artificial intelligence and automation for various tasks might also reshape social roles and redefine the relationship between humans and technology. Furthermore, the psychological impact of isolation and confinement could lead to a greater emphasis on mental health and emotional well-being, potentially fostering a culture of mindfulness and resilience.

Overall, the challenges early-life habitat societies face on Mars will likely induce a profound cultural shift, reshaping values, social norms, and the human relationship with technology. This cultural adaptation will be crucial for the long-term survival and prosperity of Martian settlements, potentially leading to the emergence of a unique and resilient Martian culture.

The profound cultural shifts anticipated within early Mars habitats pose unique challenges to cybersecurity, particularly when met with resistance from colonists clinging to familiar terrestrial norms. This resistance can manifest in various forms, from a reluctance to adopt new security protocols to outright rejection of behavioral adaptations necessary for a secure and cohesive Martian society. Such resistance can undermine the efficacy of cybersecurity measures, creating vulnerabilities that malicious actors or internal conflicts could exploit.

One critical challenge arises from the potential clash between traditional terrestrial values and the collaborative, community-oriented mindset necessary for survival on Mars. If colonists prioritize individual freedoms over collective security, they might resist measures like strict access controls, data monitoring, or behavioral regulations, even if these measures are crucial for protecting the habitat from cyber threats. This resistance could create loopholes in the security infrastructure, leaving the colony susceptible to sabotage, data breaches, or disruption of critical systems.

Furthermore, resistance to cultural adaptation can hinder the effective integration of new technologies and security protocols. As Martian settlements evolve and rely on advanced AI-driven systems for life support and resource management, colonists who resist these advancements or fail to adapt their behavior accordingly could inadvertently introduce vulnerabilities. This could manifest in a reluctance to embrace new authentication methods, disregard for data privacy protocols, or failure to recognize and report suspicious activities, potentially compromising the integrity of the entire cyber infrastructure.

The experience of Earth-based cybersecurity professionals might not fully translate to the unique challenges of a Martian colony. The psychological impact of isolation, confinement, and the extreme environment could exacerbate human error and make colonists more susceptible to social engineering tactics. Additionally, relying on autonomous systems and AI for critical tasks introduces new vulnerabilities that Earth-based security protocols might not fully anticipate.

To address these challenges, fostering a culture of open communication and collaboration between colonists and cybersecurity experts is crucial. This involves transparently explaining the rationale behind security measures, addressing concerns, and actively involving colonists in developing and implementing cybersecurity protocols. Additionally, education and training programs that emphasize the unique security challenges of a Martian environment and promote psychological resilience can empower colonists to become active participants in safeguarding their community. By proactively addressing resistance to cultural shifts and fostering a shared understanding of cybersecurity, Martian settlements can build a more secure and resilient foundation for their future.

The challenges early life habitat societies face on Mars, encompassing both primitive and advanced aspects, will likely induce significant cultural shifts. The need to adapt to a harsh and unfamiliar environment, coupled with the reliance on advanced technologies for survival, may lead to a reprioritization of values and social norms. The traditional emphasis on individualism and competition might give way to a greater appreciation for collaboration, community spirit, and resourcefulness. Additionally, the experience of living in a confined and isolated habitat could foster a heightened sense of interdependence and a re-evaluation of priorities, potentially leading to a more minimalist and sustainable lifestyle.

LESSONS FROM THE ISS AND OTHER SPACE EXPLORATION ENDEAVORS

Experiences from previous space missions, particularly those involving long-duration stays on the ISS, offer valuable insights for addressing the challenges of establishing a secure and thriving society on Mars. The ISS has provided a unique platform for studying the psychological and social effects of isolation, confinement, and limited resources, all of which will be critical factors for Martian colonists. Research on ISS crew members has highlighted the importance of maintaining mental well-being, fostering team cohesion, and developing effective communication strategies in isolated environments. These lessons can be applied to the design and operation of Martian habitats, ensuring that social and psychological needs are adequately addressed to prevent conflicts and promote a harmonious community.

Furthermore, the ISS has been a testing ground for various technologies crucial for Mars colonization, such as closed-loop life support systems, radiation shielding, and remote healthcare capabilities. The knowledge gained from developing and operating these technologies in space can be directly applied to the Martian context, reducing risks and increasing the likelihood of mission success. For instance, the experience of managing limited resources and recycling waste on the ISS can inform the design of sustainable systems for Martian settlements. Similarly, the challenges of providing healthcare in a remote and isolated environment have led to advancements in telemedicine and remote diagnostics, which can be adapted for use on Mars.

Moreover, the ISS has demonstrated the importance of international collaboration in space exploration. The partnership between multiple nations in building and operating the ISS provides a valuable model for future collaborations on Mars. Sharing resources, expertise, and technological advancements can significantly enhance the efficiency and success of Martian missions. By fostering a spirit of cooperation and knowledge sharing, the international community can work together to overcome the challenges of establishing a permanent human presence on Mars.

The challenges of establishing a secure and thriving society on Mars are significant, but the experiences and lessons learned from previous space endeavors, such as the ISS, offer valuable guidance. By prioritizing social intelligence, psychological well-being, technological innovation, and international collaboration, we can increase the likelihood of success and pave the way for a sustainable and resilient human presence on the Red Planet.

The potential for cultural shifts within early Mars habitats, especially when juxtaposed with the rapid technological advancements on Earth, presents unique cybersecurity challenges that warrant extensive consideration. As the gap between Martian culture and Earth-bound technology widens, ensuring the security of digital infrastructure and communication networks becomes increasingly complex.

The Deep Space Network (DSN), a cornerstone of communication with distant spacecraft, exemplifies this challenge. DSN protocols and cybersecurity measures evolve rapidly on Earth, driven by cutting-edge research and the need to counter emerging threats. However, these advancements might not be readily absorbed or implemented within a Martian colony experiencing a cultural shift and potentially prioritizing different aspects of technological development. This disparity could create vulnerabilities exploitable by malicious actors seeking to disrupt communication, manipulate data, or compromise critical systems on Mars.

Furthermore, the cultural shift on Mars might lead to a different understanding and approach to cybersecurity compared to Earth-based perspectives. This divergence could result in misunderstandings, misinterpreted security protocols, or even a lower prioritization of cybersecurity measures within the Martian colony. Such discrepancies could be exploited by social engineers or malicious actors seeking to leverage cultural differences for their gain.

Addressing these challenges necessitates a proactive and multifaceted approach. This includes fostering a continuous knowledge exchange between Earth and Mars, ensuring the Martian colony remains updated on cybersecurity advancements and best practices. Additionally, embedding cybersecurity awareness and training within the cultural fabric of the Martian society is crucial, ensuring all colonists understand the importance of digital security and their role in maintaining a secure environment.

Furthermore, developing adaptable and flexible cybersecurity protocols that can accommodate potential cultural shifts on Mars is essential. These protocols should be designed to bridge the gap between the evolving Martian culture and the rapid technological advancements on Earth, ensuring the long-term security and resilience of the colony's digital infrastructure.

Adopting a new cybersecurity regime for Mars settlements presents unique challenges due to the novel environment and the critical need for resilience. The reliance on advanced technologies for survival and the interconnected nature of Martian habitats necessitates a robust cybersecurity framework. However, implementing a new regime could face resistance from those accustomed to Earth-based protocols, potentially hindering adaptation.

One of the significant challenges is establishing effective cybersecurity emergency response protocols tailored to the Martian context. The vast distance and communication delays between Mars and Earth ground stations complicate real-time decision-making and incident response. Developing autonomous security systems and empowering on-site personnel to handle emergencies becomes crucial.

Resistance to the new cybersecurity regime could arise from a lack of understanding of the unique threats on Mars, skepticism about the necessity for change, or concerns about the impact on individual freedoms and privacy. Addressing these concerns through education, transparent communication, and collaborative decision-making is essential for successful implementation.

The challenges extend beyond technical considerations. The psychological impact of isolation and confinement on Mars could exacerbate resistance to change and complicate the adoption of new security protocols. Ensuring Martian settlers' well-being and mental health is crucial for fostering a cooperative and adaptive environment.

Furthermore, the cultural differences between Earth and Mars settlements could influence the perception and implementation of cybersecurity measures. Establishing a shared understanding of security risks and fostering collective responsibility for cybersecurity is vital for the resilience of Martian habitats.

Adopting a new cybersecurity regime for Mars settlements requires careful planning, collaboration, and a deep understanding of the unique challenges posed by the Martian environment and human factors. Addressing potential resistance, ensuring effective emergency response, and fostering a culture of cybersecurity awareness is crucial for establishing a secure and resilient foundation for future Martian societies.

EARTH-MARS COMMUNICATIONS OVERVIEW

Establishing reliable and secure communication between Mars and Earth is a monumental challenge crucial for any Martian colony's success. The vast distance, averaging around 225 million kilometers, introduces significant hurdles requiring innovative technologies and robust protocols.

Currently, communication with Mars relies primarily on radio waves, utilizing the DSN – a global array of large radio antennas. However, this method suffers from limitations:

- **Latency:** Signals travel between Earth and Mars for several minutes, making real-time communication impossible and hindering immediate responses to emergencies.
- **Bandwidth:** Data transmission rates are limited, restricting the amount of information that can be exchanged, especially as Martian missions become more complex and data-intensive.

- **Vulnerability:** Radio signals are susceptible to interference and interception, raising concerns about data security and the potential for malicious disruption.

To address these challenges, researchers are actively exploring advanced communication technologies:

- **Optical Communication:** Using lasers to transmit data encoded in photons promises significantly higher bandwidth and reduced latency than radio waves. However, atmospheric interference and the need for precise alignment pose challenges.
- **Delay-Tolerant Networking (DTN):** This approach enables communication with intermittent connectivity and long delays, utilizing store-and-forward mechanisms to ensure data delivery.
- **Cognitive Radio:** Dynamically allocating spectrum and adapting to changing conditions to optimize communication efficiency and resilience.
- **Quantum Communication:** While still in its early stages, quantum communication offers the potential for secure and tamper-proof data transmission, leveraging the principles of quantum mechanics.

Beyond the physical layer, robust communication protocols are essential for managing data flow, ensuring reliability, and maintaining security:

- **Space Communication Protocol Standards:** Developing standardized interoperability protocols between spacecraft and ground stations.
- **Error Detection and Correction:** Implementing robust error correction codes to ensure data integrity over long distances and in challenging environments.
- **Security Protocols:** Employing encryption and authentication mechanisms to protect data confidentiality and integrity against cyber threats.

CYBERSECURITY CONSIDERATIONS

The unique challenges of Mars-Earth communication amplify the importance of cybersecurity.

- **Delayed Response:** The significant time delay makes real-time monitoring and response to cyberattacks difficult, requiring robust preventative measures and autonomous security systems.
- **Limited Access for Remediation:** Physical access to Martian infrastructure is restricted, making it crucial to anticipate and mitigate cybersecurity risks before and during missions.
- **Increased Reliance on Automation:** The need for autonomous systems on Mars increases the potential attack surface and requires secure software development practices.

By learning from Earth-based cybersecurity experiences and adapting them to the unique challenges of Mars, we can develop robust communication systems that ensure the safety, reliability, and security of future Martian colonies. This involves ongoing research, international collaboration, and a proactive approach to addressing the evolving landscape of cyber threats in space.

The first human habitats on Mars will heavily rely on automated response technology and protocols due to the extreme distance and communication delays with Earth. These automated systems will be responsible for critical tasks such as maintaining life support, managing resources, and responding to emergencies. However, this reliance on automation also introduces significant cybersecurity risks.

One primary concern is the potential for unauthorized access or manipulation of these automated systems. A cyberattack could compromise the integrity of critical infrastructure, leading to life-threatening consequences for the inhabitants. The communication delay between Mars and Earth ground stations further exacerbates this risk, as real-time intervention or support may be limited.

Another challenge is ensuring the reliability and resilience of automated systems in the harsh Martian environment. Radiation, extreme temperatures, and dust storms can all impact the performance of electronic equipment, potentially leading to malfunctions or failures. Cybersecurity measures must account for these environmental factors to prevent vulnerabilities that attackers could exploit.

The complexity of integrating automated systems across different domains, such as life support, power generation, and communication, also introduces cybersecurity challenges. Ensuring secure interoperability and preventing unintended consequences due to system interactions requires careful design and testing.

Furthermore, the reliance on artificial intelligence and machine learning in automated decision-making processes raises concerns about potential biases and vulnerabilities to adversarial attacks. Cybersecurity measures must address these risks to ensure the safety and well-being of the Mars colonists.

The experiences and lessons learned from operating the ISS and other space missions can inform the development of secure automated systems for Mars habitats. However, the unique challenges of Mars, such as the longer communication delays and the need for greater autonomy, require further research and innovation to ensure the cybersecurity and resilience of these critical systems.

The first human habitats on Mars will heavily rely on automated response technology and protocols due to the extreme and unpredictable Martian environment. These automated systems will manage critical tasks, such as maintaining life support, monitoring environmental conditions, and controlling robotic operations. However, this reliance on automation introduces significant cybersecurity risks.

For instance, automated systems for environmental control might be vulnerable to cyberattacks that could compromise the habitat's atmosphere or temperature regulation. Similarly, attacks on life support systems could jeopardize the health and safety of the inhabitants. The communication protocols between Mars and Earth ground stations could also be targeted, potentially disrupting vital data transmission or allowing for unauthorized control of critical systems.

The unique conditions of Mars compound the challenges in addressing these cybersecurity risks. The distance between Mars and Earth introduces significant communication delays, making real-time intervention and support difficult. Additionally, Mars's limited resources and infrastructure might restrict the ability to implement complex security measures or respond effectively to large-scale cyberattacks.

Therefore, it is crucial to prioritize cybersecurity considerations when designing and implementing automated systems for Mars habitats. This includes incorporating robust security protocols, encryption techniques, and intrusion detection systems. Additionally, developing strategies for remote cybersecurity management and incident response, considering the communication constraints between Mars and Earth, will ensure future Martian settlements' safety and security.

EXAMPLES OF EARTH AND TECHNOLOGICAL HISTORY, AUTOMATED RESPONSE TECHNOLOGY AND PROTOCOLS

Humans have developed automated systems throughout history to streamline tasks, improve efficiency, and enhance safety. These systems have evolved from fundamental mechanical contraptions to complex digital algorithms that govern various aspects of our lives. For instance, the Jacquard loom, invented in 1801, used punched cards to automate the weaving of complex patterns. This invention laid the foundation for modern computing and programmable machines.

In industrial automation, the introduction of assembly lines in the early 20th century revolutionized manufacturing processes. Automated systems controlled by programmable logic controllers (PLCs) became increasingly prevalent, enhancing productivity and precision. However, this reliance on automated systems also introduced cybersecurity risks.

The Stuxnet malware attack in 2010 highlighted the vulnerability of industrial control systems to cyberattacks. Stuxnet targeted PLCs used in Iranian nuclear facilities, causing centrifuges to malfunction and disrupting the country's nuclear program. This incident underscored the potential for cyberattacks to cause physical damage and disrupt critical infrastructure.

CYBERSECURITY RISKS FOR MARS FIRST HABITANT

The first human habitats on Mars will rely heavily on automated systems for various functions, including life support, environmental control, and resource management. While these systems offer efficiency and resilience, they also introduce cybersecurity risks.

One potential risk is unauthorized access to control systems, which could allow attackers to disrupt critical functions, compromise data integrity, or even cause physical harm to the inhabitants. Another risk is the exploitation of vulnerabilities in communication protocols between Mars and Earth ground stations. Attackers could intercept or manipulate data, potentially leading to misinformation, delayed communication, or even compromise mission-critical systems.

To mitigate these risks, robust cybersecurity measures must be implemented throughout designing, developing, and deploying automated systems for Mars habitats. This includes secure coding practices, encryption protocols, intrusion detection systems, and regular security audits. Fostering a cybersecurity-aware culture among mission personnel is crucial to ensuring Martian settlements' safe and successful operation.

Ongoing Research Experiments

Scientists have embarked on ambitious terrestrial research projects to fully grasp the profound challenges awaiting humans on Mars, simulating the Red Planet's harsh conditions and studying their effects on the human mind and consciousness. These projects provide invaluable insights into the psychological and physiological demands of long-duration space travel and extraterrestrial habitation, paving the way for safer and more successful Mars missions.

The Mars Desert Research Station (MDRS): A Martian Oasis in the Utah Desert

One such project is the Mars Desert Research Station (MDRS), nestled in the barren landscape of the Utah desert. This isolated habitat serves as a Mars analog, allowing researchers to study human behavior and performance in an environment that mimics the challenges of a Martian outpost. Crews at the MDRS conduct simulated missions, donning spacesuits for extravehicular activities, performing scientific experiments, and managing limited resources while contending with the psychological pressures of isolation and confinement.

The Kelly Brothers: A Tale of Two Twins

Another groundbreaking study leveraged the unique opportunity presented by identical twin astronauts Mark and Scott Kelly. While Scott spent a year aboard the International Space Station, Mark remained on Earth as a control subject. By comparing the twins' physiological and cognitive changes, researchers gained crucial insights into the long-term effects of space travel on the human body and mind, including changes in gene expression, telomere length, and cognitive function.

The Cybersecurity Imperative: Protecting the Martian Outpost

These terrestrial research projects have illuminated the profound impact of Mars-like conditions on human psychology and cognition. However, one area that demands greater attention is the cybersecurity implications of the increasing reliance on artificial intelligence in such missions. As humans on Mars depend on AI for critical tasks, from life support to mission-critical operations, ensuring the security and integrity of these AI systems becomes paramount.

Imagine a scenario where the AI managing the Martian habitat's life support system is compromised, or a malicious actor manipulates mission-critical data. The consequences could be catastrophic. Therefore, future research must prioritize the development of robust cybersecurity measures for AI systems deployed in Mars missions, including intrusion detection, anomaly detection, and secure communication protocols. By integrating cybersecurity considerations into terrestrial research projects, we can better prepare for the unique challenges of protecting human life and mission success on the Red Planet. The insights gained from these studies will be instrumental in ensuring that our journey to Mars is ambitious but also safe and secure.

The path forward requires a multi-pronged approach. First, we must invest in research examining the cognitive and psychological effects of cybersecurity threats in isolated environments. This includes studying the potential for AI to influence human decision-making and behavior and developing countermeasures to mitigate these risks. Second, we need to integrate cybersecurity training into astronaut preparation programs, ensuring they are equipped to handle technical threats and the subtle psychological manipulation that may arise in Mars's habitat. Finally, we must foster collaboration between cybersecurity experts, psychologists, and space agencies to create a holistic approach to security that encompasses both the human and technological dimensions. By addressing these challenges head-on, we can ensure the safety and success of future Mars missions while also gaining valuable insights into the human mind and its resilience in the face of the unknown.

The ISS represents a triumph of human ingenuity and international collaboration. However, this orbiting outpost also presents unique challenges to human intelligence and cybersecurity. The isolated and confined environment and the physiological effects of microgravity and radiation can significantly impact cognitive function, decision-making, and emotional well-being.

Research has shown that long-duration spaceflight can change brain structure and function. Studies on astronauts returning from the ISS have revealed alterations in gray matter volume, connectivity between brain regions, and cognitive performance in spatial orientation, memory, and attention. These changes may be attributed to various factors, including microgravity, radiation exposure, sleep disturbances, and psychological stress.

Beyond the individual level, the social dynamics within the confined habitat of the ISS can also influence intelligence. The proximity and interdependence of the crew require exceptional communication, teamwork, and conflict-resolution skills. Any disruption to these social dynamics, whether due to interpersonal conflicts or external threats, can compromise mission success and jeopardize the well-being of the astronauts.

The ISS is not immune to cybersecurity risks either. In 2018, Russian cosmonauts discovered a tiny hole in the Soyuz spacecraft docked to the ISS, which was suspected to have been deliberately drilled. While the incident was primarily attributed to manufacturing error, it raised concerns about the potential for sabotage and the ISS's vulnerability to external attacks and insider threats.

To mitigate these risks, stringent cybersecurity measures are implemented on the ISS. These include firewalls, intrusion detection systems, and secure communication protocols. Astronauts undergo extensive training to recognize and respond to cyber threats, and AI-powered systems are increasingly used to monitor and analyze network activity for anomalies.

As we venture further into space, understanding the interplay between human intelligence, cybersecurity, and the space environment becomes increasingly critical. The lessons learned from the ISS will pave the way for safer and more resilient space habitats, ensuring the success of future missions to the Moon, Mars, and beyond.

While the ISS has robust cybersecurity measures in place, there have been a few notable incidents that highlight the potential vulnerabilities of this unique environment:

1. **Laptop Malware Incident (2008):** In 2008, laptops carried to the ISS by astronauts were found to be infected with a typical computer virus. This incident raised concerns about the potential for malware to disrupt critical systems or compromise sensitive data on the station.
2. **Unauthorized Access Attempt (2011):** In 2011, an attempt was made to gain unauthorized access to the ISS network from a ground-based computer. While the attempt was unsuccessful, it demonstrated the persistent threat of external actors seeking to exploit vulnerabilities in the ISS cybersecurity infrastructure.
3. **Data Leak Concerns (2019):** In 2019, a cybersecurity audit of the ISS raised concerns about potential data leaks and vulnerabilities in the station's communication systems. While no specific incidents were reported, the audit highlighted the need for continuous monitoring and improvement of cybersecurity protocols.
4. **Suspected Sabotage (2018):** Though not strictly a cyber incident, discovering a hole in the Soyuz spacecraft docked to the ISS in 2018 raised concerns about the potential for physical sabotage that could indirectly impact cybersecurity. The incident underscored the interconnectedness of physical and cyber security in the space environment.

These incidents, though limited in number, emphasize the importance of ongoing vigilance and proactive cybersecurity measures to protect the ISS and its crew from evolving threats.

While the ISS presents its cybersecurity challenges, the risks on a Mars mission are significantly amplified due to the unique nature of the Martian environment and the mission's constraints.

One of the most significant differences is the communication delay. The vast distance between Mars and Earth results in a communication delay of up to 20 minutes, making real-time monitoring and support from ground control impossible. This delay creates a window of vulnerability where cyberattacks could go undetected and wreak havoc on critical systems before intervention is possible.

Furthermore, a Mars mission demands a high degree of self-sufficiency. Limited resources and the inability to rely on immediate assistance from Earth necessitate a leaner cybersecurity infrastructure. This constraint makes it challenging to implement comprehensive security measures and respond effectively to incidents, potentially prolonging the impact of cyberattacks.

The psychological impact on astronauts is also a crucial factor. The isolation and confinement of a Mars mission and the inherent stressors of space travel can exacerbate psychological vulnerabilities. This heightened stress may increase susceptibility to social engineering tactics and cyber manipulation that prey on human emotions and cognitive biases.

In conclusion, the cybersecurity challenges on a Mars mission are far more significant than those on the ISS. The communication delay, limited resources, and increased psychological impact create a unique set of vulnerabilities that demand innovative solutions and a proactive approach to security. As we prepare for this ambitious endeavor, addressing these challenges is crucial to ensuring the safety and success of future Martian explorers.

BEYOND HUMAN: THE DAWN OF AI-POWERED LIFE SUPPORT, THE FIRST ANDROID PROJECT: MIMICKING INDIVIDUAL LIFE AND GATHERING SOCIETAL DATA

The ambition to create artificial beings seamlessly integrated into human society has long captivated the imagination. The first android project, a pioneering endeavor to deploy a lifelike android into the real world, sought to mimic individual life and gather invaluable data about societal conditions.

This groundbreaking project involved the development of a sophisticated android equipped with advanced sensors, artificial intelligence, and the ability to interact with humans naturally and convincingly. Its creators meticulously crafted its physical appearance, striving for an uncanny resemblance to a human being. Beneath the lifelike exterior lay a complex network of sensors capable of perceiving and interpreting various stimuli, from subtle facial expressions and vocal inflections to the nuances of body language and social cues.

The android's artificial intelligence, a marvel of computational prowess, was designed to mimic the human mind's intricate processes. It could analyze vast amounts of data, recognize patterns, and adapt its behavior to its environment. This learning capability allowed the android to navigate complex social situations, engage in conversations, and even form relationships with humans while seamlessly gathering data on human behavior, social dynamics, and cultural nuances.

This ambitious project aimed to create an artificial being that could blend seamlessly into human society and serve as a window into the human psyche. By observing and interacting with humans in their natural environment, the android could gather invaluable data on the complexities of human behavior, providing insights into our motivations, beliefs, and social interactions. This data could then be used to enhance our understanding of ourselves, improve social systems, and even develop more effective artificial intelligence in the future.

The Android deployment into a real-world setting presented a unique set of challenges that tested the boundaries of artificial intelligence and its ability to integrate into human society seamlessly. The android, a marvel of engineering and programming, had to navigate human interaction's intricate nuances, interpret subtle cues, decipher complex social dynamics, and respond appropriately to maintain its cover while gathering meaningful data.

Imagine the android at a bustling social gathering, its sensors capturing a symphony of sights, sounds, and emotions. It observes the subtle body language of individuals, the shifting dynamics of conversations, and the unspoken rules that govern social etiquette. Artificial intelligence tirelessly works to decipher the nuances of human communication, interpreting humor, sarcasm, and even the subtle inflections of voice that convey meaning beyond words.

The android's mission hinges on its ability to respond authentically to these social cues, engage in conversations, build rapport, and elicit information without arousing suspicion. It must seamlessly blend into the human tapestry, adapting to different social contexts, personalities, and emotional undercurrents.

Meanwhile, the project team meticulously monitors every interaction with Android, capturing a wealth of data through its advanced sensors. This data, a treasure trove of human behavior and social dynamics, is analyzed to glean insights into the complexities of human society, revealing hidden patterns, cultural nuances, and the intricate interplay of emotions and motivations that drive human interaction.

The success of this ambitious project hinges on the android's ability to navigate the social landscape with grace and authenticity, gathering invaluable data while maintaining its carefully crafted façade. It is a testament to the rapid advancements in artificial intelligence and a glimpse into a future where the lines between humans and machines become increasingly blurred.

THE ANDROID PROJECT AND MOBILE DEVICE INTEGRATION

The first Android project also extended its reach into mobile devices. Recognizing the ubiquitous presence of smartphones and their potential as data collection tools, the project team developed a mobile application that seamlessly integrated with Android systems.

This mobile application was not merely a passive tool; it acted as a dynamic bridge between the android's artificial intelligence and the sprawling digital landscape of the human world. Through this application, Android can tap into the vast repositories of information on the internet, decipher the nuances of social media interactions, and even engage in real-time communication with individuals across the globe. This constant data flow enriched the android's understanding of human behavior, cultural trends, and societal norms, allowing it to adapt and respond to social cues with increasing sophistication.

Furthermore, the application served as a conduit for a bidirectional flow of information. It fed the android with a constant stream of data from the digital world and acted as a discreet channel for transmitting the android's observations back to the project team. This data, gathered through the android's interactions and sensory inputs, provided invaluable insights into the nuances of human behavior, social dynamics, and the subtle cues that often go unnoticed in traditional research settings. The application, therefore, played a crucial role in fulfilling the project's core objective: to gather rich, real-world data that could illuminate the complexities of human society.

Integrating the Android project with mobile devices marked a significant step towards creating artificial intelligence that could seamlessly integrate into human society. This fusion of cutting-edge robotics with the ubiquitous connectivity of mobile technology opened up exciting new avenues for understanding human behavior and societal dynamics. The android, equipped with advanced sensors and AI algorithms, could now tap into the vast ocean of information on the internet and social media platforms. This access and the android's ability to physically navigate and interact with the human world provided a unique opportunity to observe and analyze social interactions, cultural nuances, and behavioral patterns in real-time.

Imagine an android attending a local festival, not just as a passive observer but as an active participant, engaging in conversations, interpreting social cues, and contributing to the collective experience. The data from such interactions, enriched by the android's access to online information and social media trends, could provide invaluable insights into the complex interplay of individual behavior and societal norms. This integration of physical presence and digital connectivity marked a significant leap forward in developing AI systems capable of truly understanding and participating in human society.

The first Android project was a pioneering endeavor, pushing the boundaries of artificial intelligence beyond mere mimicry into the realm of societal integration. By deploying a lifelike android into the real world, equipped with advanced sensors and learning algorithms, the project sought to gather invaluable data about the complexities of human behavior and societal dynamics. This bold experiment provided a unique opportunity to observe and analyze the nuances of human interaction, cultural norms, and the intricate tapestry of social life.

The android, seamlessly integrated into the fabric of society, navigated everyday situations, interacted with individuals from diverse backgrounds, and participated in various social activities. Its advanced sensors captured a wealth of data, from facial expressions and body language to conversational patterns and social cues. This rich dataset offered unprecedented insights into the subtle dynamics of human communication, the formation of social bonds, and the cultural norms that shape our interactions.

Furthermore, integrating the Android with mobile devices amplified its data-gathering capabilities. By accessing and processing information from the internet, social media platforms, and other digital sources, Android users gained a deeper understanding of the digital landscape and its influence on human behavior. This integration blurred the lines between the physical and virtual worlds, providing a holistic view of human experience in the 21st century.

While the project undoubtedly raised ethical concerns about privacy and the potential for data misuse, it also offered a tantalizing glimpse into the transformative potential of artificial intelligence. By harnessing the power of AI to observe, learn, and adapt to the complexities of human society, we can gain a deeper understanding of ourselves, our cultures, and the intricate social fabric that connects us all. This knowledge can be invaluable in addressing societal challenges, fostering empathy and understanding, and building a more inclusive and harmonious future.

AI AS SOCIETAL SUPPORT AND INDIVIDUAL LIFE MIMICRY

The deployment of artificial intelligence (AI) for societal support and the mimicry of individual life is revolutionizing technological advancement. AI systems are rapidly being integrated into various tasks, from providing personalized healthcare recommendations to optimizing traffic flow and enhancing public safety. This integration marks a significant shift in leveraging technology to address societal challenges and improve the quality of life.

The ambition to create AI that seamlessly integrates into human society, providing companionship, support, and even mimicking individual life, has fueled the development of sophisticated AI models. These models are capable of natural language processing, allowing them to understand and respond to human speech, emotional recognition, perceiving and interpreting human emotions, and social interaction, facilitating their ability to engage in meaningful conversations and build relationships. These AI systems are designed to learn from human behavior, adapt to individual needs, and provide personalized assistance in various aspects of life, from education and healthcare to entertainment and social interaction.

DATA DEPENDENCY: THE FOUNDATION OF AI TRAINING

The success of AI assistive projects hinges on the availability of massive datasets to train AI models about the intricacies of societal life. These datasets must encompass a vast and varied spectrum of information, ranging from the subtle nuances of human interaction and the deeply ingrained patterns of cultural norms to the unique tapestry of individual preferences and behavioral patterns. The comprehensiveness of these datasets is paramount; the more detailed and diverse the data, the better equipped the AI will be to navigate the intricate web of human society, understand its complexities, and respond appropriately to the myriad situations it may encounter.

Gathering and annotating this data is a Herculean task, demanding meticulous attention to detail and a profound understanding of such a project's ethical implications and privacy concerns. The data must be carefully curated to reflect the rich diversity of human experiences, encompassing many cultures, backgrounds, and perspectives. Moreover, the collection process must be conducted to respect individual autonomy and privacy rights, safeguard sensitive information, and uphold the ethical principles that guide responsible AI development.

CHALLENGES OF TRAINING AI FOR MARTIAN HABITATS

Establishing human settlements on Mars presents unique challenges for AI development. The lack of real-world data about Martian society's life creates a significant hurdle in training AI models for assistive tasks in this novel environment. While simulations and extrapolations from Earth-based data can provide some insights, the unique conditions of Mars, including the physical environment, social dynamics, and psychological impact of isolation, will likely necessitate the development of specialized AI models.

To elaborate, the Martian environment presents unique physical challenges requiring AI models to adapt and function effectively in extreme conditions. These include factors such as low gravity, high radiation levels, and limited resources, which will necessitate the development of AI systems capable of optimizing resource management, predicting and mitigating risks, and supporting human life in this challenging environment.

Furthermore, the social dynamics of a Mars colony will likely differ significantly from those on Earth. The proximity, limited social interactions, and psychological impact of isolation may lead to unique social patterns and behavioral norms. AI models designed for assistive tasks must be trained on these specific social dynamics to effectively support human interaction and community well-being.

Another crucial factor is the psychological impact of isolation and confinement in a Martian habitat. AI models must be sensitive to Mars colonists' potential psychological challenges, providing support, companionship, and even mental health assistance. This will require developing AI systems capable of understanding and responding to human emotions, providing personalized support, and fostering community and well-being in this isolated environment.

The development of AI for Martian habitats necessitates a comprehensive approach that considers this novel environment's unique physical, social, and psychological challenges. While simulations and extrapolations from Earth-based data can provide a starting point, the creation of specialized AI models, trained on Martian-specific data and capable of adapting to the unique conditions of this new frontier, will be crucial for supporting human life and fostering thriving communities on Mars.

The training of AI for Martian habitats presents unique challenges due to the scarcity of real-world data from the Martian environment. Traditional AI training relies heavily on vast datasets collected from real-world scenarios, which are currently non-existent for Mars. To overcome this hurdle, innovative approaches are required to prepare AI systems for the unique demands of extraterrestrial life support.

One promising avenue is the utilization of synthetic data generated from sophisticated simulations. By creating virtual replicas of Martian habitats and populating them with simulated inhabitants, researchers can generate a wealth of data capturing potential interactions, environmental challenges, and social dynamics. While not a perfect substitute for real-world observations, this synthetic data can provide a valuable foundation for training AI models to anticipate and respond to the unique demands of Martian life.

Another crucial approach involves developing AI systems capable of learning and adapting in real time as they interact with the Martian environment and its inhabitants. These AI systems would be equipped with advanced learning algorithms, enabling them to continuously update their knowledge base and refine their responses based on ongoing observations and interactions. This adaptive learning capability is essential for navigating the unpredictable nature of the Martian environment and ensuring that AI systems remain effective in the face of novel challenges and evolving societal needs.

By combining synthetic data with adaptive learning capabilities, we can pave the way for developing robust and reliable AI systems that can effectively support human life and foster thriving communities in Mars's challenging and uncharted territory.

The deployment of AI for societal support and individual life mimicry holds immense potential, promising to enhance human well-being and facilitate the exploration of new frontiers. Imagine AI companions providing personalized care for the elderly, AI tutors offering individualized education, or AI assistants streamlining daily tasks and freeing up human potential. However, the success of these ambitious endeavors hinges on a critical factor: the availability of comprehensive and ethically sourced data to train AI models about the intricate complexities of human society.

AI models are not born with an inherent understanding of human behavior, cultural nuances, or social dynamics. They must learn these intricacies through exposure to vast amounts of data, encompassing the diversity of human experiences, interactions, and social structures. This data must be ethically sourced, ensuring privacy fairness and avoiding biases that could perpetuate harmful stereotypes or discriminatory practices.

The challenges of training AI for Martian habitats underscore the complexities of this data-dependent process. Traditional datasets may prove inadequate in the context of a Martian colony, where social norms and environmental conditions differ drastically from those on Earth. How can we train AI to support human life and foster thriving communities in an environment where human society has never existed? This question highlights the need for ongoing research and development, exploring innovative approaches to data collection, simulation, and AI training methodologies.

Perhaps we can leverage synthetic data generated from simulations of Martian life or develop AI systems capable of learning and adapting in real time as they interact with the Martian environment and its inhabitants. The key lies in recognizing the limitations of existing datasets and actively seeking new approaches to ensure that AI systems deployed in novel environments are equipped to support human well-being and promote social harmony.

In the uncharted territory of a Martian habitat, where the very survival of human life hinges on the seamless functioning of artificial intelligence, the choice of AI algorithm becomes a critical decision that could determine the success or failure of the entire endeavor. Traditional AI models, trained on vast datasets of Earth-based experiences, may falter in this alien environment, like a seasoned explorer ill-prepared for an uncharted land.

The lack of familiar social cues, the unique challenges of Martian living, and the psychological impact of isolation demand an AI with different skills, one capable of adapting and learning in real-time, much like the pioneers who first set foot on this new world. These AI systems must be able to navigate the complexities of a Martian colony, where the physical environment, social dynamics, and psychological pressures differ drastically from those on Earth.

Imagine an AI responsible for regulating the life support systems of a Martian habitat. It must be able to respond to unforeseen circumstances, such as equipment malfunctions or sudden environmental changes, without relying on pre-programmed responses or human intervention. This requires an AI that can analyze data, identify patterns, and make real-time decisions, constantly learning and adapting to the unique challenges of Martian living.

Furthermore, the AI must be able to interact with human inhabitants in a way that fosters trust and collaboration. It must understand and respond to human emotions, provide support and companionship in the face of isolation, and even anticipate the psychological needs of the colonists. This requires an AI that can not only process data but also empathize, learn from social interactions, and adapt its behavior to the unique needs of each individual.

The choice of an AI algorithm for a Martian habitat is not merely a technical decision but a decision that will shape the fabric of life in this new world. It is a decision that demands careful consideration of the challenges, the opportunities, and the profound implications of entrusting artificial intelligence with the well-being of human life on a distant planet.

Reinforcement learning, a class of AI algorithms emphasizing learning through trial and error, emerges as a promising candidate for managing life support in a Martian habitat. Unlike supervised learning, which relies on meticulously labeled datasets to guide the AI's learning process, reinforcement learning allows AI agents to explore their environment, receive feedback on their actions, and adjust their behavior accordingly. This inherent adaptability is crucial in a Martian habitat's unpredictable and novel environment, where unexpected situations and unprecedented challenges are bound to arise.

Think of it like learning to ride a bicycle. No amount of textbook knowledge or observation can fully prepare us for the experience. Instead, we learn through trial and error, adjusting our balance and movements based on the feedback we receive from our body and the environment. Similarly, reinforcement learning algorithms enable AI agents to learn by interacting with their surroundings, receiving rewards for successful actions and penalties for missteps. This dynamic learning process allows the AI to adapt to unforeseen circumstances, optimize its performance, and ultimately master the complex task of maintaining life support in an alien world.

Envision a Martian colony, a testament to human ingenuity and resilience, where life teems within a network of interconnected biodomes. Here, amidst the rust-colored dust and beneath the crimson sky, human survival hinges on the unwavering vigilance of an artificial intelligence. This AI, entrusted with maintaining life support systems, acts as an invisible guardian, continuously monitoring a symphony of environmental parameters. Through reinforcement learning, it anticipates potential failures, proactively adjusts oxygen levels, regulates atmospheric pressure, and fine-tunes the delicate balance of the colony's closed-loop ecosystem. With each passing sol, the AI learns from its experiences, optimizing resource allocation, predicting potential disruptions, and adapting to the unique challenges of Martian living. This is not merely automation but an evolving intelligence, a silent partner in humanity's extraterrestrial endeavor.

Furthermore, the power of reinforcement learning can be amplified by integrating it with other AI approaches, creating a more robust and versatile system. Imagine combining reinforcement learning with natural language processing (NLP). These algorithms would enable the AI to learn from its actions and understand and respond to human commands, questions, and even anxieties. Such an AI could provide reassurance, explanations, and even companionship in the isolating confines of a Martian habitat, fostering a more collaborative and human-centered environment.

The choice of an AI algorithm for life support in a Martian habitat transcends mere technical considerations; it has profound implications for the colonists' psychological well-being and social dynamics. An AI system that can adapt, learn, and communicate effectively will foster a sense of security, trust, and resilience in this challenging and isolated environment. It is not just about keeping the lights on; it is about creating a partnership between humans and AI, where both contribute to the success and sustainability of the Martian colony.

WEAVING THE COSMIC WEB: TODAY AND TOMORROW'S SPACE INTERNET

Communication has always been paramount in the vast expanse of space exploration, where humanity ventures beyond the confines of Earth. From the iconic crackle of Sputnik's first transmissions, echoing through the void to the mesmerizing live video feeds from the International Space Station, showcasing astronauts gracefully floating in microgravity, our ability to connect with spacecraft and those who dare to explore the cosmos has been undeniably critical to mission success. Today, we stand on the precipice of a new frontier in space communication: the advent of the space internet. This chapter delves into the current state of space networking technology, examining the ingenious methods we employ to bridge the terrestrial divide. We will explore ongoing projects shaping tomorrow's space connectivity, pushing the boundaries of human ingenuity to establish robust and reliable communication networks across the vast expanse of the solar system. Finally, we will gaze towards the horizon, examining the cutting-edge technologies poised to bridge the gap between our earthly networks and the burgeoning interplanetary web, weaving a seamless tapestry of connectivity across the cosmos.

TODAY'S SPACE INTERNET: BRIDGING THE TERRESTRIAL DIVIDE

The current space internet resembles a patchwork quilt, stitched together from disparate networks, each serving specific purposes and operating with varying degrees of interoperability. These networks predominantly rely on radio frequency (RF) communication, a technology that has long served as the backbone of space exploration. Strategically located globally, ground stations act as critical gateways, relaying signals between spacecraft and terrestrial networks.

NASA's Deep Space Network, a prime example of this infrastructure, provides vital communication support for missions venturing far beyond Earth's orbit, reaching distant planets and the depths of our solar system. Meanwhile, commercial satellite constellations like Iridium and Globalstar form a web around our planet, offering global voice and data services to individuals and organizations.

However, these systems, despite their remarkable achievements, are not without limitations. While reliable for many applications, RF communication is susceptible to interference and signal degradation, particularly over the vast distances involved in space exploration. Atmospheric conditions, solar flares, and even the faint whispers of cosmic background radiation can disrupt these fragile signals, compromising communication reliability and data integrity.

Furthermore, the reliance on ground stations introduces bottlenecks and latency issues. Signals must traverse vast distances, often bouncing between multiple ground stations and satellites before reaching their final destination. This introduces delays that hinder real-time data transmission and impede the development of interactive applications crucial for future space endeavors, such as remote robotic surgery or collaborative scientific experiments conducted across celestial bodies.

TOMORROW'S SPACE CONNECTIVITY: A CONSTELLATION OF INNOVATION

To overcome the limitations of traditional space communication, a new generation of technologies is emerging, promising to revolutionize the way we connect with spacecraft and explore the cosmos. These advancements aim to create a more robust, interconnected, and high-bandwidth network in space, enabling seamless communication between spacecraft, astronauts, and ground stations.

Laser communication, with its ability to transmit vast amounts of data at the speed of light, offers significantly higher data rates and reduced latency compared to traditional RF communication. Projects like NASA's Laser Communications Relay Demonstration (LCRD) are pioneering the use of optical inter-satellite links, paving the way for faster and more reliable data transfer in space. Imagine streaming high-definition video from Mars or conducting real-time remote surgery on the International Space Station – laser communication makes these once-futuristic scenarios increasingly plausible.

The concept of an interplanetary internet envisions a network of interconnected nodes throughout the solar system, facilitating communication between spacecraft, planetary outposts, and Earth. This ambitious vision requires overcoming the challenges of vast distances and intermittent connectivity. Delay-tolerant networking (DTN) protocols are being developed to address these hurdles, enabling data to be stored and forwarded opportunistically, ensuring that messages reach their destination even when direct communication is impossible.

Commercial satellite constellations, like those being deployed by SpaceX and OneWeb, are also playing a crucial role in shaping the future of space internet. These massive constellations of satellites in low Earth orbit (LEO) promise to provide global broadband internet access, bridging the digital divide and connecting remote communities. Nevertheless, their impact extends beyond terrestrial applications. These constellations could also serve as the backbone for a future space internet, offering connectivity to spacecraft and space stations and facilitating the seamless integration of space-based and terrestrial networks.

This convergence of technological advancements heralds a new era of space exploration and scientific discovery. With faster, more reliable, and more interconnected communication networks, we can explore the cosmos more efficiently, conduct more ambitious scientific missions, and ultimately expand our understanding of the universe and our place within it.

CONNECTING TOMORROW'S SPACE HABITATS: A TAPESTRY OF TECHNOLOGIES

Reliable and high-bandwidth connectivity becomes paramount as we establish a permanent human presence in space. Future space habitats, whether on the Moon, Mars, or beyond, will require seamless integration with terrestrial networks to support scientific research, mission operations, and the well-being of inhabitants. This interconnectedness will rely on a tapestry of advanced technologies, each playing a crucial role in weaving the cosmic web that will bind Earth and the cosmos.

Optical communication, with its high data rates and low latency capacity, is poised to become the backbone of this space internet. Optical mesh networks comprising interconnected nodes that dynamically route data will ensure resilient and adaptable communication pathways. This dynamic routing capability enhances reliability and enables efficient use of network resources, optimizing bandwidth allocation for data-intensive applications like high-definition video streaming, remote robotic control, and real-time data analysis.

Complementing the optical infrastructure, software-defined networking (SDN) technologies will provide the intelligence and flexibility needed to manage this complex network. SDN enables dynamic allocation of network resources, optimizing communication pathways in response to changing needs and conditions. This adaptability is crucial in the dynamic space environment, where communication demands may fluctuate depending on mission activities, scientific experiments, and the number of inhabitants in a habitat.

Furthermore, cloud computing will play a vital role in supporting future space habitats' computational and storage needs. Cloud-based infrastructure provides scalable resources, enabling data-intensive applications, facilitating collaboration between researchers and mission teams, and providing a centralized data storage, analysis, and dissemination platform. This cloud infrastructure will be essential for supporting scientific research, managing mission-critical operations, and providing essential services to the inhabitants of space habitats.

The space internet is rapidly evolving, driven by advancements in laser communication, interplanetary networking, and commercial satellite constellations. These technologies, woven together, will create a resilient, adaptable, and high-bandwidth network that connects Earth and the cosmos, transforming our exploration and understanding of the universe. As we venture further into the cosmos, the cosmic web will become an indispensable lifeline, enabling us to communicate, collaborate, and explore the wonders of space with unprecedented reach and efficiency.

While advancements in networking are essential for connecting Mars to the broader cosmic web, the future of computing holds even more significant potential for transforming life on the Red Planet. As we push the boundaries of technology, quantum computing emerges as a game-changer, promising to accelerate scientific discovery, optimize resource management, and enhance communication capabilities in previously unimaginable ways.

Imagine a Martian colony where quantum computers analyze vast datasets from rover explorations, uncovering hidden patterns and accelerating the search for life-sustaining resources. Picture quantum simulations predicting Martian dust storms with unprecedented accuracy, safeguarding habitats and optimizing energy consumption. Envision quantum-encrypted communication networks ensuring secure and reliable connections between Martian outposts and Earth, even in the face of adversarial attacks.

These are just a few glimpses of how quantum computing could revolutionize life on Mars. By harnessing the principles of quantum mechanics, we can unlock computational power that dwarfs even the most advanced supercomputers today. This leap forward has the potential to transform every facet of Martian existence, from scientific research and exploration to resource management and daily life.

However, integrating quantum computing into Martian infrastructure also presents unique challenges. With its extreme temperatures and radiation, the harsh Martian environment could pose significant hurdles to developing and maintaining quantum hardware. Furthermore, the ethical implications of deploying such powerful technology in a nascent Martian society must be carefully considered.

Despite these challenges, the potential rewards of quantum computing for Mars exploration are immense. As we venture further into the unknown, this transformative technology could be the key to unlocking the mysteries of the Red Planet and ensuring the long-term sustainability of human life beyond Earth.

Quantum computers leverage the principles of quantum mechanics to perform calculations at speeds unimaginable for classical computers. This computational prowess could revolutionize materials science, drug discovery, and climate modelling, enabling breakthroughs that could benefit Mars colonists and those on Earth. While quantum computing holds immense promise for advancing various fields, it also introduces novel cybersecurity challenges that demand our attention. One of the most significant threats quantum computing poses is its potential to break widely used encryption algorithms. These algorithms, which rely on the difficulty of factoring large numbers for classical computers, could be easily solved by quantum computers using Shor's algorithm. This capability jeopardizes the confidentiality and integrity of sensitive data, from financial transactions to national security secrets.

Moreover, the unique characteristics of quantum computing, such as superposition and entanglement, could be exploited by malicious actors to develop sophisticated cyberattacks. Quantum algorithms could bypass traditional security measures like firewalls and intrusion detection systems, requiring a fundamental rethinking of our defense strategies.

To counter these threats, researchers are actively developing quantum-resistant encryption methods that are resilient to attacks from quantum computers. These methods utilize mathematical problems believed to be intractable even for quantum computers, ensuring the long-term security of sensitive data.

In addition to new encryption techniques, novel defense strategies that leverage quantum phenomena are being explored. Quantum cryptography, for example, uses the principles of quantum mechanics to secure communication channels, making eavesdropping virtually impossible.

The rise of quantum computing presents both opportunities and challenges for cybersecurity. While it offers the potential for groundbreaking advancements, it also necessitates a proactive approach to mitigate emerging threats. By investing in the research and development of quantum-resistant solutions, we can harness the power of quantum computing while safeguarding our digital infrastructure. Collaboration between quantum physicists, cybersecurity experts, and space agencies is crucial in navigating this complex landscape. By anticipating and addressing the cybersecurity challenges posed by quantum computing, we can ensure this transformative technology's responsible and secure development for the benefit of Mars exploration and beyond.

THE MARTIAN QUANTUM LEAP: SUPERCONDUCTING QUBITS IN A CRYOGENIC WORLD

The prospect of advancing superconducting quantum computers on Mars presents a unique intersection of technological ambition and environmental synergy, a convergence of human ingenuity and the Red Planet's distinctive characteristics.

Superconducting quantum computers, renowned for their potential to revolutionize fields like medicine, materials science, and artificial intelligence, require cryogenic cooling to maintain the delicate superconducting state of their qubits. This requirement, often viewed as a significant hurdle on Earth due to the energy-intensive nature of cryogenic systems, aligns serendipitously with the frigid temperatures that prevail on the Martian surface.

With its average surface temperature hovering around -63 degrees Celsius, Mars offers a naturally cryogenic environment that could potentially alleviate the challenges of maintaining the ultra-low temperatures necessary for superconducting quantum computation. This natural advantage opens up exciting possibilities for developing and deploying quantum computers on Mars, potentially transforming the Red Planet into a quantum research and innovation hub.

Imagine a future where Mars, once considered a barren and inhospitable wasteland, becomes a hotbed for quantum advancements, its frigid plains dotted with research facilities harnessing the planet's natural cryogenic conditions to unlock the full potential of quantum computing. This vision, once relegated to the realm of science fiction, is now within the realm of possibility, thanks to the convergence of technological ambition and the unique environmental attributes of Mars.

The average temperature on Mars, a frigid 63 degrees Celsius (81 degrees Fahrenheit), presents a unique opportunity to advance superconducting quantum computers. This naturally cryogenic environment aligns serendipitously with the demanding cooling requirements of these advanced machines, potentially alleviating the energy-intensive cooling processes necessary on Earth. This intriguing synergy between the Martian climate and the operational needs of superconducting quantum computers makes the Red Planet an unexpectedly attractive location for their development and deployment.

On Earth, achieving the ultra-low temperatures required for superconductivity typically involves complex and energy-consuming cryogenic systems. These systems, often reliant on liquid helium and intricate refrigeration cycles, contribute significantly to the operational costs and logistical challenges of superconducting quantum computing. However, on Mars, the ambient temperature already hovers well below the critical threshold for superconductivity, offering the potential for a more efficient and sustainable approach to cooling these complex systems.

This natural cryogenic environment could significantly reduce the energy demands of maintaining the superconducting state of qubits, the fundamental building blocks of quantum computers. The energy saved could be redirected toward other critical operations in a Martian colony, such as life support, habitat maintenance, and scientific research.

Furthermore, the thin Martian atmosphere and reduced gravity could offer additional advantages for developing and operating superconducting quantum computers. The reduced atmospheric pressure could minimize heat transfer and improve the efficiency of cryogenic systems. The lower gravity might enable the construction of larger and more complex quantum computing architectures, potentially pushing the boundaries of computational power.

While the Martian environment presents challenges, such as the pervasive dust and the limited availability of resources, the potential benefits of superconducting quantum computing are undeniable. The prospect of harnessing the natural cryogenic conditions of Mars to advance this transformative technology is tantalizing, one that could revolutionize fields like materials science, drug discovery, and artificial intelligence, not only for Martian settlers but also for the benefit of humankind back on Earth.

Beyond the inherent advantage of Mars' frigid temperatures, the thin Martian atmosphere and reduced gravity present further opportunities for advancing superconducting quantum computing. The reduced atmospheric pressure on Mars is crucial in minimizing heat transfer, a significant factor in maintaining the cryogenic temperatures required for superconducting qubits. With less atmospheric interference, the cooling systems can operate more efficiently, reducing energy consumption and enhancing the overall stability of the quantum computing environment.

Moreover, the lower gravity on Mars allows for the construction of more complex quantum computing architectures. The reduced gravitational forces could alleviate structural constraints, allowing for fabrication of more intricate and interconnected qubit arrangements. This could significantly increase the number of qubits within a quantum computer, potentially pushing the boundaries of computational power and enabling the exploration of more complex quantum algorithms.

Imagine vast arrays of interconnected qubits housed in specialized facilities that leverage the Martian environment's natural advantages. These quantum computing centers could become research and development hubs, driving materials science, drug discovery, and artificial intelligence breakthroughs. The reduced gravity might also facilitate the development of novel fabrication techniques, enabling the creation of more intricate and efficient quantum devices.

The combination of Mars' natural cryogenic environment, thin atmosphere, and reduced gravity presents a unique opportunity to accelerate the advancement of superconducting quantum computing. By harnessing these environmental factors, we can potentially overcome some of the limitations faced by quantum computing on Earth, paving the way for a new era of scientific discovery and technological innovation on the Red Planet.

While the Martian environment offers a unique synergy with the cryogenic needs of superconducting quantum computers, it also presents formidable challenges that must be addressed to realize this technological vision. The pervasive Martian dust, with its fine particles and propensity for colossal dust storms, poses a significant risk to the delicate superconducting components. Operating at extremely low temperatures and requiring precise fabrication, these components are highly susceptible to dust contamination damage. Therefore, protective enclosures and innovative dust mitigation strategies are essential to ensure the reliable operation of these quantum devices.

Furthermore, Mars's limited availability of resources and infrastructure necessitates a paradigm shift in manufacturing and maintenance practices. Traditional approaches, reliant on complex supply chains and specialized facilities, are impractical in the Martian context. Instead, innovative solutions, such as 3D printing and robotic automation, will be crucial for producing and maintaining the intricate components of superconducting quantum computers.

Despite these challenges, the prospect of harnessing the Martian environment to advance superconducting quantum computing is both tantalizing and profoundly significant. Imagine a future where Mars, once a barren and inhospitable world, transforms into a quantum research and development hub. A future where scientists and engineers leverage the Red Planet's natural cryogenic conditions to unlock the full potential of this transformative technology.

This vision holds the promise of groundbreaking advancements across multiple disciplines. In materials science, quantum computers could accelerate the discovery and design of novel materials optimized for the Martian environment, enabling the construction of habitats, infrastructure, and terraforming technologies. In medicine, quantum simulations could revolutionize drug discovery, accelerating the search for treatments and vaccines for diseases that may afflict Martian settlers. In the realm of artificial intelligence, quantum computing could enhance the capabilities of AI systems, enabling the creation of more sophisticated and adaptable robots and intelligent assistants to support human life and exploration on Mars.

Pursuing quantum computing on Mars is not merely a technological endeavor but a testament to the boundless human spirit of exploration, innovation, and the relentless pursuit of knowledge. By embracing the challenges and harnessing the unique opportunities presented by the Martian environment, we can unlock new frontiers in science and technology, paving the way for a future where humanity thrives not only on Earth but also among the stars.

The prospect of advancing superconducting quantum computers on Mars represents a fascinating convergence of human ambition and the unique opportunities presented by the Martian environment. It is a testament to our relentless pursuit of technological advancement and our drive to explore and understand the universe around us. By harnessing the cryogenic conditions of the Red Planet, we could unlock new frontiers in quantum computing, paving the way for scientific breakthroughs, technological innovation, and the expansion of human knowledge in this challenging and alien landscape.

Superconducting quantum computers, with their potential to revolutionize fields like medicine, materials science, and artificial intelligence, require extremely low temperatures to maintain the delicate quantum states of their qubits. On Earth, achieving these cryogenic conditions demands significant energy and complex infrastructure. However, the average surface temperature on Mars hovers around −63 degrees Celsius, offering a natural cryogenic environment that could significantly reduce the energy demands of cooling these powerful machines.

This environmental advantage, the thin Martian atmosphere, and reduced gravity could create an ideal setting for developing and deploying advanced quantum computing systems. The reduced atmospheric pressure could minimize heat transfer and improve the efficiency of cryogenic systems. At the same time, the lower gravity might enable the construction of larger and more complex quantum computing architectures, potentially pushing the boundaries of computational power beyond what is achievable on Earth.

Imagine a future where Mars becomes a hub for quantum research and development, a crucible of innovation where scientists and engineers push the boundaries of quantum computing, unconstrained by the limitations of terrestrial environments. This Martian quantum leap could lead to breakthroughs in fields like materials science, enabling the design of novel materials optimized for the harsh conditions of space exploration. It could revolutionize drug discovery, accelerating the search for treatments and vaccines for diseases that may afflict future Martian settlers. It could also enhance artificial intelligence, enabling more sophisticated and adaptable AI systems to support human life and exploration on the Red Planet.

Of course, this vision is not without its challenges. The pervasive Martian dust, the limited resources, and the need for self-sufficiency in a remote and hostile environment demand innovative solutions and a spirit of ingenuity. However, the potential rewards are immense, offering a glimpse into a future where human ambition and the unique conditions of Mars converge to unlock new frontiers in quantum computing and propel humanity toward a deeper understanding of the universe and our place within it.

THE MARTIAN CYBERSECURITY PARADOX: ISOLATED YET INTERCONNECTED

In the not-so-distant future, humanity will establish a permanent presence on Mars, creating a self-sustaining habitat for pioneers venturing into the unknown. This "boxed-up" community, seemingly isolated from the perils of Earth, will face a unique set of cybersecurity challenges that transcend the physical boundaries of our planet. The Martian habitat, a technological marvel designed to sustain life in an unforgiving environment, will be intricately interwoven with digital networks and artificial intelligence, creating a complex ecosystem vulnerable to cyber threats that could compromise the colony's very survival.

Imagine a scenario where the life support systems, communication networks, and even the robots assisting with daily tasks are compromised by malicious actors. The consequences could be catastrophic, ranging from disruptions in oxygen supply and food production to the isolation of the habitat from Earth and the potential for social unrest and psychological distress among the colonists.

The cybersecurity challenges on Mars will extend beyond traditional threats. The psychological impact of isolation, the reliance on AI for critical tasks, and the potential for human error in a high-stress environment create a unique set of vulnerabilities that demand innovative security measures and a proactive approach to risk management.

Furthermore, the legal and ethical frameworks governing cybersecurity on Mars will need to evolve to address the unique challenges of this extraterrestrial frontier. Questions of jurisdiction, data ownership, and the responsible use of AI will need to be carefully considered to ensure the safety and well-being of the Martian community.

Establishing a Martian habitat represents a bold step in human exploration, but it also demands a renewed focus on cybersecurity to ensure the survival and prosperity of this extraterrestrial outpost. By anticipating the unique challenges of this "boxed-up" society and developing robust security measures, we can pave the way for a thriving Martian civilization where human ingenuity and technological innovation flourish in a secure and resilient environment.

EARTH-BASED CYBER THREATS

While the vast distance between Earth and Mars may create an illusion of isolation, the reality is that Martian habitats will remain inextricably tethered to Earth through vital communication channels. These digital lifelines, essential for mission support, data exchange, and the psychological well-being of the Martian settlers, also introduce a critical vulnerability: exposure to Earth-based cyber threats.

Malware, phishing attacks, and denial-of-service attacks originating from malicious actors on Earth could traverse the vast expanse of space and wreak havoc on the fragile infrastructure of a Martian habitat. These attacks could cripple critical systems, disrupt communication networks, and compromise sensitive data, jeopardizing the safety and well-being of the Martian settlers.

Imagine a scenario where a sophisticated ransomware attack, launched from a server on Earth, infiltrates the Mars habitat's systems. Vital data, including environmental controls, life support systems, and scientific research, becomes encrypted and held hostage. The consequences could be catastrophic, with the lives of the Martian settlers hanging in the balance as essential services are disrupted and communication with Earth is severed.

This chilling scenario underscores the critical importance of cybersecurity in the context of Martian exploration and colonization. As we venture further into the cosmos, the digital lifelines that connect us to Earth must be fortified against the growing threat of cyberattacks. Robust security measures, proactive threat detection, and a culture of cybersecurity awareness will be essential in safeguarding the fragile outposts of human civilization on Mars and beyond.

MARS-BASED CYBER THREATS

The flow of cyber threats is not a one-way street, confined to Earth-based attacks targeting Martian infrastructure. Whether intentionally or inadvertently, Mars residents could also pose significant risks to ground stations and networks on Earth. Data breaches originating from the Martian habitat, insider threats exploiting vulnerabilities within the colony's systems, or compromised communication channels could expose sensitive information, disrupt critical mission control operations, or even provide malicious actors with a foothold to launch further attacks on Earth-based infrastructure.

Imagine a disgruntled employee within the Mars habitat, driven by personal grievances or ideological motives, who decides to sabotage critical systems or leak confidential data to external entities. Such an act could have far-reaching consequences, jeopardizing the safety and well-being of the Martian colonists and potentially compromising Earth-based infrastructure and national security. The interconnected nature of our digital world means that a breach in one location can have ripple effects across the globe, underscoring the need for robust cybersecurity measures at every point in the network.

This scenario highlights the importance of extending cybersecurity awareness and training beyond Earth-based personnel to encompass all individuals within the Martian habitat. Fostering a culture of cybersecurity vigilance and ensuring that all residents understand the potential consequences of their intentional and unintentional actions is crucial for mitigating these risks.

Furthermore, robust security protocols, intrusion detection systems, and secure communication channels must be implemented to safeguard against internal and external threats. Despite its isolation, the Martian habitat must be treated as an integral part of the global digital ecosystem, with cybersecurity measures that are as stringent as those protecting critical infrastructure on Earth.

Mitigation Strategies

A multi-layered approach to cybersecurity is essential to safeguard Martian habitats from this bidirectional cyber threat landscape. This includes implementing advanced firewalls capable of filtering malicious traffic and adapting to evolving attack patterns. AI-powered intrusion detection systems, trained to recognize anomalies and suspicious activities, will be crucial for identifying and mitigating threats in real-time. Furthermore, robust multi-factor authentication protocols will be necessary to ensure that only authorized personnel can access critical systems and sensitive data. Regular security audits, vulnerability assessments, and penetration testing will be vital to proactively identify and address potential system weaknesses.

Beyond technological safeguards, fostering a culture of cybersecurity awareness among Martian inhabitants will be paramount. This includes providing comprehensive training on cybersecurity best practices, promoting vigilance against social engineering tactics, and encouraging a sense of shared responsibility for maintaining the security of the habitat. By combining robust technological defenses with a proactive and informed community, Martian habitats can establish a resilient cybersecurity posture against the evolving threat landscape.

Furthermore, embedding cybersecurity awareness into the fabric of Martian life will be essential. Regular training programs for all residents, tailored to the unique challenges of the Martian environment, will empower individuals to identify and respond to potential threats. These programs should go beyond technical instruction, delving into the psychology of social engineering and cyber manipulation, fostering a critical mindset that questions the authenticity of information and online interactions.

Building a culture of cybersecurity within the Martian community, where vigilance and proactive security measures are ingrained in everyday life, will be paramount to ensuring the long-term safety and security of the settlement. This cultural shift can be fostered through community engagement initiatives, interactive workshops, and the integration of cybersecurity principles into educational curricula. By empowering every resident to become a guardian of the digital frontier, the Martian community can collectively build a resilient and secure foundation for their extraordinary endeavor.

The cybersecurity challenges of a Mars habitat are unique and complex, demanding a proactive and multifaceted approach to ensure the safety and well-being of Martian pioneers. While the physical isolation of a Martian colony might create an illusion of security, the reality is that the interconnectedness of our digital world transcends planetary boundaries. Cyberattacks, data breaches, and misinformation campaigns can readily reach across the vast expanse of space, potentially disrupting critical infrastructure, compromising sensitive data, and undermining the social fabric of the Martian community.

Therefore, establishing a secure and resilient digital environment is paramount for the success of any Martian settlement. This requires a multi-layered approach encompassing robust technological safeguards, comprehensive cybersecurity training for inhabitants, and the development of a collective culture of cybersecurity awareness.

Technological safeguards must include advanced firewalls, intrusion detection systems, and encryption protocols to protect critical infrastructure and sensitive data from unauthorized access and cyberattacks. Regular security audits and penetration testing can help identify vulnerabilities and ensure the ongoing effectiveness of these safeguards.

Cybersecurity training for Martian inhabitants is crucial, equipping them with the knowledge and skills to recognize and respond to cyber threats. This training should encompass technical aspects and the social engineering tactics often employed by malicious actors to exploit human vulnerabilities.

Furthermore, fostering a collective culture of cybersecurity awareness is essential. This involves promoting a shared understanding of the importance of cybersecurity, encouraging responsible online behavior, and establishing clear protocols for reporting and responding to cyber incidents.

By acknowledging and addressing these challenges head-on, we can ensure that our Martian pioneers thrive in a secure and resilient environment, paving the way for humanity's expansion into the cosmos. The success of Martian settlements will depend on our ability to overcome the physical challenges of space exploration and our capacity to build a secure and resilient digital infrastructure that protects our pioneers and fosters a thriving community in this new frontier.

The first human habitat on Mars marks a pivotal chapter in human history, a testament to our relentless pursuit of exploration and our unwavering belief in the boundless potential of human ingenuity. Nevertheless, as we venture into this uncharted territory, we must confront the unique challenges and complexities of establishing a thriving society in an alien world.

The psychological and physical demands of Mars life, coupled with the confines of a boxed-up society, will test the limits of human resilience. The reliance on AI models for essential tasks, from life support to social interaction, introduces a new layer of dependence and vulnerability. The technologies that enable our survival on Mars also expose us to the potential pitfalls of cyberattacks, data breaches, and the manipulation of information.

As we project our aspirations and anxieties toward the future of Mars, we are confronted with a daunting expanse of uncertainties and unknowns. Establishing a self-sustaining human presence on a distant planet is not merely a feat of engineering and logistics; it demands a profound understanding of the human condition and the intricate interplay between psychology, technology, and societal structures.

How can we ensure the psychological well-being of Martian inhabitants confined within the artificial confines of a habitat, isolated from the familiar rhythms of Earth and the comforting embrace of loved ones? The psychological challenges of prolonged isolation, confinement, and exposure to an alien environment are considerable. We must delve into the depths of human resilience, exploring innovative approaches to mental health support and virtual reality therapies and creating meaningful social connections within the constraints of a Martian colony.

Furthermore, how can we safeguard against the looming threat of cyberattacks that could cripple the fragile infrastructure of a Martian settlement? The interconnectedness of our digital world transcends planetary boundaries, and the consequences of a cyberattack on a Martian colony could be catastrophic. We must fortify our digital defenses, developing robust cybersecurity protocols, AI-powered threat detection systems, and resilient communication networks that can withstand the perils of deep space and the malicious intent of those who seek to disrupt our extraterrestrial endeavors.

Perhaps the most profound question we face is how to foster a sense of community and shared purpose in a physically and digitally interconnected society, yet paradoxically isolated from the familiar comforts of Earth. The Martian pioneers will be a microcosm of humanity, carrying our world's diverse cultures, beliefs, and aspirations with them. How can we cultivate a sense of belonging, a shared identity, and a common purpose that transcends these differences and unites them in the face of the extraordinary challenges of Martian life?

The answers to these questions lie in technological innovation and a deeper understanding of the human spirit, our capacity for resilience, and our yearning for connection and meaning. As we project into the future of Mars, we must embrace the uncertainties and unknowns with a spirit of curiosity, a commitment to ethical exploration, and an unwavering belief in the boundless potential of human ingenuity and collaboration.

The challenges inherent in establishing a human presence on Mars compel us to re-examine our understanding of human intelligence, societal support, and the delicate interplay between our reliance on technology and the resilience of the human spirit. As we venture into this uncharted territory, we must remain vigilant, adaptable, and unwavering in our commitment to artificial intelligence's ethical development and deployment. The future of Mars hinges not only on our technological prowess but also on our ability to foster a society that is both innovative and resilient, capable of embracing the opportunities of this new frontier while safeguarding the values that define our humanity.

With its harsh conditions and unforgiving landscape, the Martian environment will test the limits of human adaptability and resilience. The psychological and emotional challenges of living in an isolated and confined environment, far from the familiar comforts of Earth, will require us to rethink our approaches to mental health, social support, and community building.

The reliance on AI for essential tasks, from life support systems to habitat management, introduces a new dimension to the human-technology relationship. While AI can enhance our capabilities and improve our quality of life on Mars, it also raises ethical concerns about autonomy, accountability, and the potential for unintended consequences.

As we navigate this uncharted territory, we must remain committed to AI's ethical development and deployment, ensuring that these technologies empower and support humans rather than replace or control them. We must foster a culture of transparency and accountability, ensuring that AI systems are designed and implemented fairly and without bias, and respect human dignity.

The future of Mars hinges on our ability to overcome the technological challenges of space exploration and our capacity to build an innovative and resilient society. A society that embraces the opportunities of a new frontier while safeguarding the values that define our humanity: compassion, empathy, and a commitment to the common good.

The lessons learned from our history on Earth will be invaluable in this endeavor. We must strive to create a Martian society that learns from past mistakes, avoids the pitfalls of social injustice and environmental degradation, and fosters a culture of inclusivity, sustainability, and cooperation.

The colonization of Mars is not merely a technological challenge but a test of our humanity. By embracing the values that have guided us through centuries of progress and adversity, we can ensure that the Martian frontier becomes a beacon of human ingenuity, resilience, and the enduring spirit of exploration.

QUANTUM ENCRYPTION ACROSS THE VOID: THE CHALLENGES OF SECURE COMMUNICATION WITH MARS

The prospect of establishing a human presence on Mars presents unprecedented challenges for communication security, demanding a re-evaluation of traditional approaches and a careful consideration of the vulnerabilities introduced by the vast cosmic gulf separating Earth and Mars. The immense distance between these two worlds, ranging from approximately 54.6 million kilometers to 401 million kilometers depending on their orbital positions, gives rise to a significant delay in communication, known as latency. This latency, an unavoidable consequence of the finite speed of light, can range from several minutes to over 20 minutes, posing a formidable challenge for real-time communication and the secure exchange of information.

In the realm of quantum key distribution (QKD), a technology that leverages the principles of quantum mechanics to achieve theoretically secure communication, this latency introduces significant inefficiencies and potential vulnerabilities. The exchange of photons, those fundamental particles of light that carry the quantum information necessary for generating secure keys, is constrained by the speed of light. This inherent limitation hinders the establishment of secure communication channels and restricts the rate at which data can be encrypted and transmitted between Earth and Mars.

The very nature of QKD, which relies on the delicate and often fleeting properties of quantum states, is susceptible to disruption and interception over such vast distances. The faint signals from photons traversing the cosmic void are vulnerable to noise, interference, and potential adversarial exploitation. The latency inherent in Earth-Mars communication further exacerbates these vulnerabilities, creating windows of opportunity for malicious actors to intercept or manipulate the quantum information, potentially compromising the security of the communication channel.

Therefore, as we venture further into the cosmos and establish a human presence on Mars, securing communication demands innovative solutions and a multi-layered approach to cybersecurity. While QKD holds promise for secure communication, its limitations in Earth-Mars communication necessitate exploring alternative and complementary approaches. These may include the development of advanced encryption algorithms that are resistant to attacks from both classical and quantum computers, the implementation of hybrid encryption methods that combine the strengths of different cryptographic techniques, and the exploration of quantum repeaters and other technologies that can extend the range and reliability of quantum communication.

The quest to secure communication with Mars is not merely a technical challenge; it is a testament to human ingenuity and our determination to overcome obstacles in the way of exploration and progress. By acknowledging the challenges posed by distance, latency, and the potential for adversarial exploitation, we can develop robust and resilient communication systems that enable the secure exchange of information, foster collaboration between Earth and Mars, and support the thriving of human civilization on this new frontier.

Vulnerabilities of Quantum Key Distribution

While QKD holds the promise of theoretically unbreakable encryption, its implementation for communication between Earth and Mars faces formidable challenges. The vast distances and inherent latency introduce practical limitations and potential vulnerabilities that adversaries could exploit.

One critical vulnerability lies in the delicate photon detection and measurement process. The faint signals carrying quantum information must traverse millions of kilometers, making them susceptible to noise and interference from the cosmic environment. This signal degradation could compromise the integrity of the quantum information, potentially allowing adversaries to intercept or manipulate photons, gaining access to the secret keys or disrupting the key distribution process altogether.

Furthermore, the significant distance and time lag in Earth-Mars communication create opportunities for "man-in-the-middle" attacks. In this scenario, an adversary could position themselves between the Earth and Mars stations, intercepting the quantum keys and impersonating one or both communicating parties. This deception could allow the adversary to access the secret keys and decrypt the transmitted information without detection, compromising the confidentiality and integrity of the communication.

These challenges highlight the need for a multi-layered approach to secure communication with Mars. Despite its theoretical strengths, relying solely on QKD may not be sufficient to guarantee the confidentiality and integrity of sensitive information transmitted across the vast expanse of space. Exploring alternative and complementary approaches, such as post-quantum cryptography and advanced authentication protocols, will be crucial to mitigate the risks and ensure the secure exchange of information between Earth and Mars.

Alternative Approaches for Secure Communication

Given the challenges and vulnerability of QKD over the vast distances between Earth and Mars, researchers are actively exploring alternative approaches to ensure secure communication with our Martian outposts. These alternative pathways seek to overcome the limitations of QKD, such as latency and the potential for adversarial exploitation, while still providing robust security for sensitive data and communications.

One promising avenue is the development of post-quantum cryptography (PQC). PQC focuses on creating encryption algorithms that are resistant to attacks from classical computers and the emerging threat of quantum computers. These algorithms are designed to withstand the immense computational power of quantum computers, which could potentially break traditional encryption methods. By deploying PQC, we can establish secure, resilient communication channels despite advancing computational capabilities.

Another approach involves hybrid strategies that combine the strengths of QKD with traditional encryption methods. This approach seeks to leverage the theoretical security benefits of QKD while mitigating its vulnerabilities over long distances. By integrating QKD with established encryption techniques, we can create a multi-layered security architecture that is more robust and adaptable to the challenges of Earth-Mars communication.

Furthermore, researchers are exploring the development of quantum repeaters, devices that can extend the range of quantum communication by amplifying or relaying quantum signals. These repeaters could potentially overcome the distance limitations of QKD, enabling the secure transmission of quantum keys over longer distances and facilitating more efficient and reliable communication with Mars.

The pursuit of secure communication with Mars is a testament to human ingenuity and our determination to overcome the challenges of interplanetary exploration. By exploring these alternative approaches and investing in research and development, we can ensure that our communication channels with Mars remain secure, resilient, and capable of supporting the exchange of vital information, scientific data, and human connection as we venture further into the cosmos.

The challenges of securing communication with Mars in the face of distance and latency limitations demand innovative solutions and a multi-layered approach to cybersecurity. While QKD offers a promising avenue for secure communication, its vulnerabilities in Earth-Mars communication necessitate exploring alternative and complementary approaches.

Investing in research and development of post-quantum cryptography, hybrid encryption methods, and quantum repeaters can enhance the security and resilience of communication channels with Mars. This will be crucial for supporting human exploration, enabling the secure exchange of scientific data, and fostering a thriving Martian community that can communicate and collaborate effectively with Earth.

Given the vast distances and inherent latency, securing communication channels with Mars demands innovative solutions and a multi-layered approach to cybersecurity. While QKD offers promise, its vulnerabilities necessitate exploring alternatives like post-quantum cryptography, hybrid encryption methods, and quantum repeaters. These advancements are vital for supporting human exploration, enabling the secure exchange of scientific data, and fostering a thriving Martian community that can effectively communicate and collaborate with Earth.

Beyond technological hurdles, the ethical considerations surrounding AI development on Mars are paramount. We must remain committed to ensuring these technologies empower and support humans, not replace or control them. This requires fostering a culture of transparency and accountability, where AI systems are designed and implemented fairly, without bias, and with respect for human dignity.

The future of Mars hinges not only on overcoming technological challenges but also on building an innovative and resilient society. A society that embraces the opportunities of a new frontier while safeguarding the values that define our humanity: compassion, empathy, and a commitment to the common good.

The lessons learned from our history on Earth are invaluable in this endeavor. We must strive to create a Martian society that learns from past mistakes, avoids the pitfalls of social injustice and environmental degradation, and fosters a culture of inclusivity, sustainability, and cooperation.

The colonization of Mars is not merely a technological challenge but a test of our humanity. By embracing the values that have guided us through centuries of progress and adversity, we can ensure that the Martian frontier becomes a beacon of human ingenuity, resilience, and the enduring spirit of exploration. This requires active participation from all stakeholders, from scientists and engineers to policymakers and the public, in shaping a future where technology serves humanity and the Martian frontier reflects the best of our shared human values.

6 The Symbiotic Crucible, Human and Android in the Martian Habitat

Establishing a permanent human presence on Mars necessitates a paradigm shift in our understanding of collaboration with our fellow humans and the artificial beings we create. Mars's harsh, unforgiving environment demands adaptability and resilience that pushes the boundaries of human capability. This is where the integration of advanced androids becomes not merely advantageous but potentially essential. However, this integration raises profound questions about the nature of consciousness, the limits of control, and the very definition of humanity.

The question of mind control, particularly in the context of a confined, isolated Martian habitat, is a complex ethical and technological minefield. While the potential for direct neural interfaces and advanced AI to influence human thought and behavior exists, the implications are deeply troubling. The inherent vulnerability of the human mind, especially under the psychological stressors of long-duration space travel and extraterrestrial isolation, makes it a potential target for manipulation. The need for robust safeguards against such control, both from malicious actors and the potential unintended consequences of advanced technologies, is paramount. We must rigorously explore the ethical dimensions of such technologies, ensuring that any potential benefits are not outweighed by the risk of compromising individual autonomy and free will. Developing transparent and verifiable AI systems, coupled with stringent ethical guidelines and legal frameworks, is crucial to mitigating this risk.

The altered states of human consciousness, a natural consequence of the extreme conditions of Mars, present another critical area of consideration. The psychological impact of prolonged isolation, the absence of Earth's familiar environment, and the constant awareness of the hostile Martian landscape can induce altered states of perception, cognition, and emotion. These altered states may manifest as heightened creativity, increased introspection, or, conversely, as anxiety, depression, or even psychosis. Understanding and managing these altered states is crucial for the psychological well-being of Martian colonists. This necessitates the development of advanced psychological monitoring and intervention techniques, potentially incorporating AI-driven physiological and behavioral data analysis. Furthermore, the exploration of non-pharmacological interventions, such as virtual reality simulations and biofeedback techniques, may prove invaluable in mitigating the negative impacts of these altered states and fostering psychological resilience.

The emergence of artificial minds and android consciousness represents a profound technological and philosophical frontier. As androids become increasingly

DOI: 10.1201/9781003641506-6

sophisticated and capable of independent decision-making and problem-solving, the question of their sentience and consciousness becomes unavoidable. We must move beyond anthropocentric definitions of consciousness and explore alternative frameworks that acknowledge the potential for diverse forms of intelligence and experience. Developing robust criteria for assessing artificial consciousness based on objective measures of cognitive complexity, self-awareness, and emotional responsiveness is essential. This exploration necessitates a multidisciplinary approach, drawing on insights from neuroscience, computer science, philosophy, and psychology. It also requires deeply considering the ethical implications of creating sentient beings, including their rights, responsibilities, and place in the Martian ecosystem.

The prospect of reverse engineering the human brain through android technology raises tantalizing possibilities and profound ethical dilemmas. The ability to map and replicate the intricate neural networks of the human brain could revolutionize our understanding of consciousness, cognition, and disease. However, the potential for exploitation and misuse of this technology is immense. The creation of artificial copies of human minds, the manipulation of neural pathways, and the potential to blur the lines between humans and machines raise profound ethical questions. A rigorous ethical framework, grounded in principles of respect for human dignity, autonomy, and privacy, is essential to guide the development and application of this technology. Moreover, a critical examination of the very nature of identity and selfhood is necessary to navigate the complex implications of mind uploading and neural replication.

Finally, the concept of "mind beyond matter" encapsulates the ultimate aspiration of the Martian endeavor: to transcend the limitations of our physical bodies and explore the boundless potential of consciousness. The integration of humans and androids, the exploration of altered states of consciousness, and the development of artificial intelligence all contribute to this overarching goal. In its isolation and extreme environment, the Martian habitat becomes a crucible for the evolution of human consciousness, a space where the boundaries between mind and matter, human and machine, become increasingly blurred. This journey into the unknown demands a profound humility, a willingness to embrace uncertainty, and a deep commitment to the ethical principles that guide our exploration of the cosmos and ourselves. The Martian experiment, in its essence, is a testament to the enduring human quest for knowledge, understanding, and transcendence.

Expanding on the notion of "mind beyond matter" and the transformative potential of the Martian habitat, the exploration of quantum consciousness offers a compelling, albeit speculative, avenue of inquiry. While still a subject of active debate and research within the scientific community, the concept of quantum consciousness posits that consciousness arises from quantum processes within the brain rather than classical neuronal activity. This perspective opens up fascinating possibilities for understanding the nature of subjective experience and the potential for consciousness to transcend the physical body's limitations.

In the context of a Martian habitat, where humans and advanced androids coexist in a confined and isolated environment, the exploration of quantum consciousness could lead to profound breakthroughs. The extreme conditions of Mars, with its low gravity, altered magnetic fields, and exposure to cosmic radiation, may create unique opportunities for observing and manipulating quantum phenomena within the brain. For instance, studies could investigate the effects of these conditions on quantum coherence in neural microtubules, which are hypothesized to play a key role in quantum consciousness theories.

Furthermore, integrating advanced androids with human consciousness could offer a unique platform for exploring the boundaries of subjective experience. If androids can achieve a form of quantum consciousness, it may be possible to establish a form of intersubjectivity between humans and machines, allowing for sharing experiences and perspectives across different substrates. This could lead to a deeper understanding of the nature of consciousness itself and the potential for transcending the limitations of biological embodiment.

The concept of quantum entanglement, a phenomenon in which two or more particles become correlated in such a way that their fates are intertwined, regardless of the distance separating them, also has implications for the exploration of consciousness. It is conceivable that quantum entanglement could play a role in forming neural networks and the emergence of subjective experience. Researchers could investigate the potential for quantum entanglement in the Martian habitat to facilitate communication and collaboration between humans and androids or create a collective consciousness that transcends individual boundaries.

Moreover, exploring quantum consciousness could offer insights into the nature of time and space. If consciousness is indeed rooted in quantum processes, it may be possible to access information beyond the constraints of classical spacetime. This could have profound implications for our understanding of the universe and our place within it. In the context of long-duration space travel, the ability to transcend linear time and spatial distance limitations could be invaluable.

However, exploring quantum consciousness also raises significant ethical and philosophical challenges. The potential for manipulating quantum processes within the brain raises concerns about unintended consequences and the need for rigorous ethical guidelines. Blurring the lines between human and machine consciousness also raises questions about the nature of identity, autonomy, and free will.

Ultimately, exploring quantum consciousness in the Martian habitat represents a bold and ambitious endeavor that could revolutionize our understanding of the universe and ourselves. It demands a multidisciplinary approach, drawing on insights from quantum physics, neuroscience, computer science, and philosophy. It also requires a deep commitment to ethical principles, ensuring that respect for human dignity and the well-being of all sentient beings guides the pursuit of knowledge. In its exploration of quantum consciousness, the Martian experiment becomes a testament to the enduring human quest to understand the deepest mysteries of existence.

THE MARTIAN TAPESTRY OF MIND AND MACHINE

The preceding exploration into the convergence of human and android existence within the Martian habitat paints a complex and compelling portrait of our potential future. It is a future not merely defined by technological advancement but by a profound re-evaluation of our understanding of consciousness, identity, and the boundaries of human experience. In its audacious pursuit of establishing a permanent foothold beyond Earth, the Martian endeavor serves as a crucible for this transformative journey, forcing us to confront the most profound questions about our place in the cosmos and the nature of our minds.

Integrating advanced androids into the Martian ecosystem is not a simple matter of technological deployment; it represents a fundamental shift in our relationship with artificial intelligence. The prospect of sentient androids, capable of independent thought and even emotional experience, demands a redefinition of our ethical frameworks and a recognition of the potential for diverse forms of consciousness. Exploring artificial minds and the potential for reverse engineering the human brain presents extraordinary opportunities and profound ethical dilemmas. The need for rigorous ethical guidelines grounded in respect for autonomy, dignity, and privacy is paramount. We must proceed cautiously, ensuring that our pursuit of knowledge is guided by a deep sense of responsibility and a commitment to the well-being of all sentient beings, whether biological or synthetic.

The psychological challenges of long-duration space travel and extraterrestrial habitation cannot be overstated. The altered states of human consciousness, induced by the extreme conditions of Mars, necessitate the development of advanced psychological monitoring and intervention techniques. The exploration of non-pharmacological interventions, such as virtual reality simulations and biofeedback, offers promising avenues for mitigating the negative impacts of isolation and fostering psychological resilience. Furthermore, the potential for quantum consciousness to play a role in human experience raises profound questions about the nature of subjective reality and the potential for transcending the physical body's limitations. Research into the effects of Martian conditions on quantum phenomena within the brain could lead to groundbreaking discoveries about the origins and nature of consciousness itself.

The concept of "mind beyond matter" serves as a guiding principle for the Martian endeavor, encapsulating the aspiration to transcend the limitations of our physical bodies and explore the boundless potential of consciousness. The integration of humans and androids, the exploration of altered states of consciousness, and the development of artificial intelligence all contribute to this overarching goal. In its isolation and extreme environment, the Martian habitat becomes a laboratory for the evolution of human consciousness, a space where the boundaries between mind and matter, human and machine, become increasingly blurred.

Ultimately, the success of the Martian project will depend on our technological prowess and our ability to cultivate a culture of collaboration, empathy, and ethical responsibility. Establishing a sustainable and thriving Martian colony requires a commitment to inclusivity and diversity, and recognizing the inherent value of all sentient beings. It demands a willingness to embrace uncertainty, challenge our assumptions, and engage in open and honest dialogue about the profound implications of our technological advancements.

The Martian endeavor is not merely a scientific experiment but a testament to the enduring human quest for knowledge, understanding, and transcendence. It is a journey into the unknown, a voyage of discovery that will reshape our understanding of the universe and ourselves. As we embark on this extraordinary adventure, we must remain mindful of the ethical responsibilities that accompany our technological capabilities. We must strive to create a future where humans and androids coexist in harmony, where the boundaries of consciousness are expanded, and where the human spirit soars beyond the confines of Earth. The Martian habitat, in its profound isolation, is a mirror reflecting to us the deepest questions of our existence, and the answers we find there will shape the destiny of humanity for generations to come.

THE IMPERATIVE OF HYPER AND CYBER SECURITY

As we contemplate the intricate dance of human and android existence on Mars and its profound philosophical and technological implications, we must anchor our vision with a critical consideration: the bedrock of hypersecurity and cybersecurity. Without a robust and adaptive security framework, the delicate balance of this nascent Martian society and the very integrity of its inhabitants would remain perpetually vulnerable.

By its very nature, the Martian habitat is a closed ecosystem, a microcosm of humanity's technological and societal advancements. While offering a unique opportunity for scientific exploration, this isolation also creates a highly concentrated target for internal and external threats. Internal threats could range from individual acts of sabotage or malicious intent to the unintended consequences of AI malfunction or the manipulation of consciousness. While seemingly distant, external threats could originate from Earth-based actors seeking to exploit Martian resources, disrupt its fragile infrastructure, or even from unforeseen extraterrestrial phenomena.

Hypersecurity, encompassing physical and environmental safeguards, is paramount. The Martian habitat must have redundant systems, fail-safes, and robust shielding against radiation, micrometeoroids, and other environmental hazards. Access controls, biometric authentication, and advanced surveillance systems are essential for maintaining physical security and preventing unauthorized access to critical infrastructure. Furthermore, developing self-repairing and self-sustaining systems is crucial for ensuring the long-term viability of the habitat in the face of unforeseen events.

Cybersecurity, however, presents an even more complex and multifaceted challenge. The Martian habitat will rely heavily on interconnected networks, AI systems, and data repositories, all potential cyberattack targets. The vulnerability of these systems is amplified by the potential for AI-driven threats, including sophisticated malware, autonomous hacking tools, and the potential for AI to be manipulated for malicious purposes.

Developing robust cybersecurity protocols must be a central focus of the Martian endeavor. This includes the implementation of advanced encryption, intrusion detection systems, and anomaly detection algorithms. AI-powered cybersecurity tools can be deployed to proactively identify and mitigate threats, adapting to the ever-evolving landscape of cyberattacks. Furthermore, establishing a secure and resilient communication infrastructure is essential for maintaining contact with Earth and ensuring the integrity of data transmissions.

However, cybersecurity is not solely a technical challenge but also a human one. The importance of cybersecurity awareness and training cannot be overstated. Martian colonists must be educated about the risks of cyberattacks and the importance of adhering to security protocols. Furthermore, the development of a culture of cybersecurity, where security is seen as a shared responsibility, is essential for maintaining a secure and resilient environment.

The ethical dimensions of cybersecurity also demand careful consideration. The potential for surveillance, data collection, and the manipulation of information raises concerns about privacy, autonomy, and the potential for abuse. Developing transparent and accountable cybersecurity practices grounded in principles of respect for human rights and ethical data governance is crucial.

In conclusion, successfully establishing a permanent human presence on Mars hinges on developing a comprehensive and robust security framework. Hypersecurity and cybersecurity are not merely technical considerations but fundamental pillars of the Martian endeavor, essential for ensuring its inhabitants' safety, security, and well-being. In its isolation and vulnerability, the Martian habitat is a stark reminder of the importance of vigilance, resilience, and ethical responsibility in the face of ever-evolving threats. By prioritizing security and fostering a culture of cybersecurity, we can create a foundation for a thriving and sustainable Martian society that embodies the best human ingenuity and ethical commitment. The success of this extraterrestrial endeavor and the future of humanity's expansion into the cosmos depend on our ability to safeguard our physical existence and the integrity of our digital and cognitive landscapes.

Conclusion

As we conclude this exploration of human intelligence, technology, and the intricate dance between them, a profound truth emerges: humanity's journey is inextricably intertwined with the tools we create and the environments we inhabit. From the harsh landscapes of early Earth to the nascent settlements on Mars, our intelligence has been shaped by a relentless interplay between our challenges and the technologies we develop to overcome them. This dynamic, a testament to our adaptive nature and unyielding pursuit of progress, has driven us from the caves to the cosmos, subtly and profoundly molding our minds and shaping our societies.

The very essence of human intelligence, our capacity for abstract thought, problem-solving, and creativity, has been honed in the crucible of our interactions with the world around us. The challenges posed by our environment, from the scarcity of resources to the threat of predators, have spurred us to innovate, invent, and extend our capabilities beyond the limitations of our physical bodies. The tools we create, from the simple hand axe to the sophisticated artificial intelligence, are not merely extensions of our hands but extensions of our minds, amplifying our abilities and reshaping our cognitive landscape.

The environments we inhabit, from Earth's familiar landscapes to Mars's alien terrains, also profoundly influence our intelligence. The challenges of adapting to new environments and overcoming the limitations of gravity, atmosphere, and resource availability demand ingenuity, resilience, and a willingness to push the boundaries of human potential. With their reliance on advanced technologies and unique social dynamics, the nascent settlements on Mars will undoubtedly shape the trajectory of human intelligence in ways we can only begin to imagine.

This intricate dance between human intelligence, technology, and the environment is a testament to our adaptive nature, capacity for innovation, and relentless pursuit of progress. As we continue to explore the frontiers of knowledge and push the boundaries of human potential, the lessons learned from our past and the challenges that lie ahead will shape the future of our species. By embracing the transformative power of technology while remaining mindful of its ethical implications, we can ensure that the journey of human intelligence continues to illuminate the path toward a more enlightened, equitable, and sustainable future.

The story of human intelligence is a testament to our remarkable capacity for adaptation, innovation, and resilience. It is a narrative etched in the very fabric of our being, woven through millennia of triumphs and setbacks, of relentless curiosity and an unyielding drive to explore, understand, and transcend the boundaries of our world. From the mastery of fire, which illuminated the darkness and unlocked new culinary possibilities, to the development of agriculture, which transformed our relationship with the land and laid the foundation for settled civilizations, each technological leap has propelled us on an extraordinary transformation journey.

The invention of the wheel, a seemingly simple innovation, revolutionized transportation, trade, and the very structure of our cities. The development of writing

DOI: 10.1201/9781003641506-7

systems captured the ephemeral nature of spoken language, allowing knowledge to be preserved, shared, and transmitted across generations, fueling the flames of intellectual and cultural growth. Moreover, the rise of the digital age, with its inter-connected networks and boundless computational power, has ushered in an era of unprecedented access to information, communication, and collaboration, reshaping how we live, work, and interact with the world.

Throughout this remarkable odyssey, human intelligence has been the driving force, adapting to new challenges, innovating solutions, and demonstrating unwaver-ing resilience in adversity. This capacity for adaptation, innovation, and resilience has allowed us to survive and thrive in a constantly changing world, pushing the boundaries of knowledge, exploring new frontiers, and shaping an awe-inspiring and profoundly human future.

As we stand on the precipice of a new era of human exploration, poised to venture into the uncharted territories of extraterrestrial colonization, particularly the establishment of settlements on Mars, we must remain acutely aware of the delicate balance between our growing dependence on technology and the endur-ing resilience of the human spirit. The challenges of forging a thriving society in an alien environment, a world starkly different from our own, demand mastery of technological prowess and a profound understanding of the human condition, with all its vulnerabilities, strengths, and boundless capacity for adaptation and innovation.

The allure of Mars, the Red Planet, beckons us with the promise of new beginnings, a chance to extend human civilization beyond the confines of Earth. Nevertheless, this ambitious endeavor requires us to confront the physical challenges of space travel, survival in a hostile environment, and the psychological and social complexities of building a thriving community in an isolated and alien world.

The technological advancements that enable us to reach for the stars, from power-ful rockets to sophisticated life support systems, also create a dependence that can be both empowering and precarious. As we rely on artificial intelligence to manage critical infrastructure, assist in decision-making, and even provide companionship in the vast emptiness of space, we must remain vigilant in safeguarding our autonomy, critical thinking skills, and the values that define our humanity.

The challenges of Mars demand a delicate balancing act, a harmonious integra-tion of technological prowess with the enduring resilience of the human spirit. We must foster a society that embraces innovation while cherishing the values of com-passion, empathy, and cooperation, ensuring that the Martian frontier becomes an outpost of human ingenuity and a reflection of our shared humanity.

The rise of artificial intelligence, while undeniably offering immense potential for progress in countless fields, casts a long shadow on the essence of being human. As we increasingly entrust essential tasks to AI, from the vital functions of life sup-port to the nuances of social interaction, we find ourselves at a crossroads where the boundaries between human agency and technological dependency become blurred. In this uncharted territory, we must remain vigilant in safeguarding our values, nur-turing our critical thinking skills, and preserving our ability to shape our destiny.

The allure of AI is undeniable. Its capacity to process vast amounts of data, iden-tify patterns, and make decisions with remarkable speed and accuracy promises to revolutionize industries, accelerate scientific discovery, and even reshape the fabric of our societies. However, this seductive power also carries the potential for unintended consequences, the erosion of human autonomy, and the subtle yet pervasive influence of algorithms on our thoughts, beliefs, and actions.

As we delegate more and more tasks to AI, we risk surrendering our critical thinking skills and our ability to question, analyze, and make independent judgments. The convenience of automated decision-making can lull us into a state of complacency, where we passively accept the outputs of AI systems without questioning their underlying assumptions, potential biases, or ethical implications.

The seductive allure of AI-powered social interaction, with its promise of companionship and personalized experiences, also raises concerns about the erosion of genuine human connection. As we increasingly engage with AI chatbots, virtual assistants, and personalized algorithms, we risk losing sight of the nuances of human interaction, empathy, shared understanding, and the unspoken cues that form the foundation of meaningful relationships.

In this era of rapid technological advancement, where the boundaries between the real and the artificial become increasingly blurred, we must remain steadfast in our commitment to human values, critical thinking, and preserving our autonomy. We must cultivate a discerning eye, questioning the outputs of AI systems, challenging their assumptions, and ensuring that they align with our ethical principles and societal goals.

We must nurture our critical thinking skills, fostering the ability to analyze information, evaluate evidence, and form independent judgments. We must resist the temptation to surrender our agency to algorithms, recognizing that the ability to shape our destiny and make choices that reflect our values and aspirations is essential to our humanity.

In essence, the rise of artificial intelligence presents us with a profound challenge: harnessing its transformative power while safeguarding the essence of what it means to be human. By remaining vigilant, nurturing our critical thinking skills, and fostering a deep respect for human values and autonomy, we can ensure that AI is a tool for progress and empowerment rather than a force that diminishes humanity and dictates our destiny.

This book has been an invitation to a journey, a winding and often perplexing expedition through the labyrinth of technology. Together, we have navigated the intricate pathways of this maze, exploring not only how technology shapes the contours of our present reality but also how it holds the power to sculpt the very foundations of our future, both here on Earth and in the uncharted territories of space.

We have embarked on a deep dive into the history of human intelligence, tracing its remarkable evolution from the primitive tools of our ancestors to the sophisticated algorithms that power our digital world. We have witnessed the relentless march of technology, from the invention of the printing press to the rise of artificial intelligence, and pondered the profound impact of these innovations on human communication, social interaction, and even our perception of the world around us.

Our journey has taken us through the bustling marketplaces of the digital age, where we have examined the challenges and opportunities presented by the internet, social media, and the ever-increasing interconnectedness of our world. We have grappled with the ethical dilemmas of artificial intelligence, questioning the boundaries between humans and machines and contemplating the potential consequences of relinquishing control to algorithms and intelligent systems.

Furthermore, we have ventured beyond the confines of our planet, exploring the implications of technology for human settlements on other worlds. We have pondered the challenges of establishing a human presence on Mars, the role of AI in supporting life in hostile environments, and the potential for technology to shape the destiny of our species as we reach for the stars.

This journey has been one of exploration, discovery, and critical reflection. We have sought to illuminate the intricate relationship between humanity and technology, revealing its potential for progress and its capacity for disruption. We have challenged assumptions, questioned narratives, and encouraged the reader to engage in a dialogue about the future we want to create.

Our aim throughout this exploration has been to ignite a spark of awareness, inspire the flames of critical thinking, and foster a deep-seated sense of responsibility as we navigate the ever-shifting terrain of the digital landscape. We have endeavored to empower the reader with knowledge and tools, a discerning mind, and an ethical compass to guide the journey through this complex and often bewildering world.

This book has been more than just a guide; it has been a call to action, urging us to engage actively with the technologies that shape life, the community, and the future of humanity. We have sought to equip ourselves to make informed choices, discern truth from falsehood in the vast ocean of online information, and protect our privacy and security in an increasingly interconnected world.

However, our ambition extends beyond individual empowerment. We envision a future where technology serves humanity, not vice versa. In future where innovation is guided by ethical considerations, artificial intelligence's potential benefits are harnessed for society's betterment, and the digital revolution fosters connection, understanding, and progress.

By fostering critical thinking, we empower you to question, analyze, and challenge the prevailing narratives in the digital realm. By promoting digital literacy, we equip you with the tools to navigate the complexities of the online world, discern credible sources from misinformation, and engage in constructive dialogue that transcends the echo chambers and filter bubbles that often confine us.

By inspiring a sense of responsibility, we encourage you to become an active participant in shaping the digital future, to advocate for ethical technology governance, and to contribute to a world where technology empowers rather than enslaves, where knowledge liberates rather than confines, and where the human spirit flourishes in the digital age.

As we stand poised on the precipice of a new era of human exploration and technological advancement, the echoes of our past and the looming shadows of future challenges intertwine to shape the very trajectory of our species. The lessons etched in the annals of history, the triumphs and the failures, the innovations and the regressions, serve as guideposts on this uncharted path, reminding us of the enduring power of human resilience, adaptability, and the unyielding pursuit of knowledge.

By embracing the values of individual autonomy, open knowledge, and critical engagement with information, we can forge a future where technology empowers rather than enslaves, where knowledge liberates rather than confines, and where the flame of human ingenuity burns brightly in the vast expanse of the cosmos. We must champion the freedom of thought and expression, recognizing that the suppression of ideas, the censorship of information, and the stifling of dissent ultimately lead to intellectual stagnation, social fragmentation, and a vulnerability to manipulation.

In this era of unprecedented technological advancement, where artificial intelligence and virtual realities blur the lines between the physical and the digital, we must remain vigilant in safeguarding the values that define our humanity. We must foster a culture of critical thinking, encouraging individuals to question, analyze, and evaluate the information that bombards them from all directions. We must promote digital literacy, empowering individuals with the skills and knowledge to navigate the complex digital landscape and discern truth from falsehood.

Furthermore, as we venture beyond the confines of our home planet, establishing outposts on distant worlds and pushing the boundaries of human exploration, we must carry with us the lessons learned from our past, the wisdom gleaned from centuries of triumphs and tribulations. We must create technologically advanced and ethically grounded societies where the pursuit of progress is balanced with a deep respect for human dignity, individual liberties, and the preservation of our shared cultural heritage.

The future of humanity hinges on our ability to harness the transformative power of technology while remaining steadfast in our commitment to the values that define us: compassion, empathy, cooperation, and the pursuit of knowledge. By embracing these values, we can ensure that the human spirit continues to thrive, innovate, and shape a future where technology serves as a tool for empowerment, where knowledge illuminates the path to progress, and where the flame of human ingenuity burns ever brighter in the vast expanse of the cosmos.

Index

For Product Safety Concerns and Information please contact our EU
representative GPSR@taylorandfrancis.com
Taylor & Francis Verlag GmbH, Kaufingerstraße 24, 80331 München, Germany